MW01490695

EQUUS RISING

RISING

How the Horse Shaped U.S. History

JULIA SOPLOP

Illustrated by Robert Spannring

 Hill Press

Pittsboro, North Carolina

Published by Hill Press, LLC.

For permission questions relating to the text or photography or for wholesale inquiries, contact: julia@juliasoplop.com. For permission questions related to the pen illustrations, contact: robert@robertspannring.com.

Library of Congress Control Number: 2020909662

ISBN: 978-1-7351113-0-8

DEDICATION

For Cricket, Nora, and Piper,
always and in everything

CONTENTS

AUTHOR'S NOTE

Much of my history education growing up was traditional: dates, battles, victors. As far as I remember, we learned about each event in history in one distinct unit at a time. We took a test on the material—dates, battles, victors—and then moved to the next unit. I don't remember making connections between topics or stopping to discuss how one event led to another. I don't remember learning until my senior year of high school that all authors inject their biases into the histories they write, that their own life experiences, cultural norms, and social statuses play into what they see fit to record and pass along to future generations.

If you think of the past as a food web—made up of many, many, interconnected food chains—studying one historical event in isolation would be like studying just one animal in an entire food web. Studying how one event contributed to the next and the next from a single perspective—say, from the victor's—would be like studying just one food chain within a food web.

My goal in writing this book was to add another accessible food chain to the food web of U.S. history. This book is not a scholarly tome brimming with newly uncovered primary sources. Instead, it's a compilation of existing knowledge, reorganized and simplified to serve as one example of how we can view the past through innumerable lenses—not just the limited perspective of a traditional history textbook.

From this angle of U.S. history, the horse provides a narrative thread that draws connections between events we often learn about in isolation from one another. This lens also naturally offers a more inclusive history of the country: one that doesn't ignore the energy source that powered the U.S. for centuries; one that includes women; one that includes people of color.

The eastern and western regions of the country developed differently and were rather disparate for much of history. I examined East and West separately through the 1800s, alternating the narrative between the two regions to show how their varying histories unfolded simultaneously. (A timeline in the back of the book will help you connect events of the two regions chronologically.) Once the transcontinental railroad, telegraph, and automobile unified the vast country by the early 1900s, I brought the narrative together.

I hope this book leaves you with a greater understanding

on two levels: one, that the horse played crucial and ever-changing roles in the founding and development of the U.S.; and two, that we can always broaden and enrich our views by shifting the lens to include additional perspectives.

Julia Soplop
Chapel Hill, NC
January 1, 2020

INTRODUCTION

E mtech galloped down the final stretch of the eighth race of the 2019 season at California's famed Santa Anita Park. Up for grabs: a $40,000 purse for the winning horse. The Thoroughbred colt had won his previous race just two weeks before, pulling in $25,000. But as he flew toward this finish, his two front legs abruptly snapped. He stumbled and fell hard, throwing his jockey. Track staff rushed to raise a screen around him to hide the disturbing scene of his thrashing body from the throng of onlookers. The track veterinarian, determining his injuries were life-ending, euthanized him on the spot. The champion Emtech was dead at age three.[1]

He was the thirty-second horse to die on the Santa Anita track in nine months, mostly from limb injuries. While Emtech's necropsy results have not been released, the horse had tested positive for a small amount of a banned painkiller, phenylbutazone, a month earlier.[2] The substance is considered a performance-enhancing drug, and his trainer later received a $1,500 fine for the violation—a slap on the wrist for what is considered a minor infraction amid a rapidly growing doping scandal that has embroiled the racing industry.

Inhabitants of what is now the United States have relied more heavily on the horse than any other animal they've domesticated. These days, horse stories that make the news are often dramatic, high-profile controversies relating to Thoroughbred racing or wild horse management policies. But the horse was not a bit player in the development of the nation; rather, it stands intricately and securely woven into our national consciousness.

Like humans, horses are social beings. In the wild, they live in family bands consisting of dominant and subordinate members. Mares assist in the care of foals that are not their own. Stallions defend mares and offspring within their bands. Horses signal to each other when danger approaches and form bonds between individual animals. The social capacities of horses also make them suited to form strong connections with humans, allowing for profound partnerships.

Gleaning the emotional power of the horse from recorded history can be difficult, but equine literature reveals the strength of the horse-human bond.

Writer Washington Irving put the cowboy legend of the pacing white stallion to paper for the first time in his 1835 chronicle, *A Tour on the Prairies*.[3] This campfire story, narrated countless times with endless variations throughout western history, told of a stunning wild stallion that attracted the attention of humans but always eluded capture. The white stallion became synonymous with the American ideal of freedom, an emblem of the harsh life on the frontier, of the right and choice to make a go of it on your own without tethers. This story, and the image of the wild horse as the representation of liberty, resonated with Americans during the era of westward expansion.

In 1877, British author Anna Sewell published the bestselling novel *Black Beauty*, a commentary on animal welfare in the age of horse-powered mechanization. It gained widespread popularity, and, tugging at heartstrings, contributed to movements demanding better treatment of horses.[4]

In 1902, Owen Wister's novel, *The Virginian*, not only defined the western genre but revolutionized the image of the cowboy as a high-class, chivalrous, ethical, and courageous figure.[5] This romantic persona, and the relationship between a cowboy and his horse, propelled the western genre of literature into existence at a time when the frontier was evaporating and new technology was pushing horses out of the lives of urban populations.

Western literature and film peaked in popularity in the 1950s. It wove horses, wild and domestic, into the fabric of American popular culture at a time when, in reality, they had mostly disappeared from it.

Marguerite Henry's 1966 novel, *Mustang, Wild Spirit of the West*, influenced the future of America's wild horses. The book injected an emotional appeal into politics and helped to garner support for wild horses' legal protection.

From solidifying the wild horse as a symbol of American freedom, to romanticizing the cowboy and his loyal steed, to shedding light on animal cruelty, to striking up political support for wild horses, equine literature reveals not only the social dynamics between horses and humans, but also how evoking those emotional connections has created cultural attitude shifts.

Yet equine literature really only hints at the horse's broader role in U.S. history as one of the greatest influences on the country's development. Whoever could most effectively wield the horse throughout history became the victor, whether victory meant land, money, status, or all three.

U nsurprisingly, the horse, which evolved across the North American Great Plains over the course of millions of years, thrived once again when Spanish *conquistadores* reintroduced it in the West, where it became indispensable to exploration and development.

Little did the Spanish realize, however, that their greatest source of strength, the horse, would become their greatest weakness when it fell into their enemies' hands. The Native American acquisition, mastery, and spread of horses in the West changed the course of the continent's history. For nearly two hundred years after the arrival of the Spanish, the Horse Nations—the native plains peoples who acquired horses—continued to rule the swath of land in the middle of what is now the U.S., despite maps showing other ownership claims.

It took settlers in the East a bit longer to figure out how to use the horse efficiently for their varying purposes and environments. Once they did, they wasted no time in selectively breeding the Spanish horses they obtained from Native Americans with the few horses they brought from their own countries of origin to create an array of differentiated American breeds.

As the territory of the U.S. continued to creep westward following the Revolutionary War, the explorers, trappers, pioneers, ranchers, farmers, and miners heading west relied on horses to develop the newly acquired land.

To link states that were once-separate colonies, Americans cut roads and built canals with the specific purpose of making them efficient for horses to haul goods and people. The construction of railroads increased the demand for horses to move wares and passengers to and from stations. The railroad also transformed the West, spurring the heyday of the cowboy of the open range, the formalized removal of Native Americans from their ancestral lands, the fencing off of new homesteads, and the capture and removal of wild horses.

In the decades following the Civil War, horses powered city life in the Northeast, as well as new, large-scale agricultural practices. The adoption of the automobile, however, caused demand for the horse to plummet. The consequences of the plunging horse population affected the economy enough to contribute to the Great Depression.

Although the horse is no longer a primary energy source, the horse industry still makes a significant contribution to the U.S. economy today. But the value of the horse is no longer a matter of consensus.

Declining demand for domestic horses as work animals means the worth and perception of wild horses have fallen, too. Bitter fights over land use and the proper way to manage wild herds continue today.

The modern U.S. Thoroughbred racing industry is mired in controversy. Claims of rampant drug use and animal cruelty abound and may finally force the industry to make significant regulatory changes to stay afloat.

Through all that humans have done to the horse, through the selective breeding and stealing and working and warring and racing, it has remained a steadfast and loyal companion, so much so that today many Americans keep horses purely to experience the strength of the horse-human bond, simply for the joy of it. Despite drastic technological advances and societal shifts that rendered the horse obsolete in most industries over time, the horse remains firmly embedded in our national imagination, an animal of vast historical and cultural influence deserving to have its story told.

HORSE TERMS TO KNOW BEFORE DIGGING IN

Equine—Relating to the horse.

Equus—The genus under which the following modern animals are classified: the horse (domestic and wild), zebras, wild asses (and the domestic donkey), the onager, and Przewalski's horse.

Equus caballus—The modern horse species, which encompasses all breeds, domestic and wild. Abbreviated *E. caballus*. (Exception: Przewalski's horse, which is a separate species.)

Foal—A juvenile horse, usually under one year old. A female foal is a filly. A male foal is a colt.

Gelding—A male horse that has been castrated to prevent reproduction.

Mare—An adult female horse.

Stallion—An adult male horse that has not been castrated.

Hands—The traditional measurement of a horse from the ground to the middle of the withers (shoulders). Long before most people had access to measuring tools, they used the width of the hand turned sideways to estimate the size of a horse. Today hands are still used to express the size of a horse, but a hand has been standardized to mean four inches. Additional inches are placed after a decimal point. For example, a horse measuring 15 hands and three inches would be denoted as 15.3 hands.

Breed—A type of horse with a particular set of characteristics as defined and registered by a breed organization. These characteristics may include lineage, size, coloration, conformation (body shape), and physical abilities. Since breed registries in the U.S. only date back to the late 1800s, I refer to horses as types or varieties, instead of breeds, prior to that time. The word breed is *not* synonymous with the word species; again, all modern horses fall under the species *E. caballus*.

Cold blood—A draft horse that is large, slow, and relatively calm. Used throughout history to carry heavily armored knights, and for agricultural work and heavy hauling. (The term cold blood does not suggest a reptile; this designation describes the horses' calm temperament.)

Hot blood—A horse that is fast and high-strung, such as a Thoroughbred. A hot blood makes a good race horse.

Warm blood—A horse that has a mix of cold blood and hot blood genes and temperament, such as the Spanish horse and today's Quarter Horse. A warm blood is a versatile animal for work or pleasure riding.

Gaits—The different leg rhythms horses use to move, meaning the ways their legs coordinate with each other. The standard gaits are the walk, trot, and canter. (Some people consider the gallop a fourth gait, while others consider it a fast canter.) But some horses are able to perform other leg rhythms, either naturally or through training, such as pacing or ambling. I refer to some horses as "gaited" if their ability to perform gaits other than the standard ones is relevant. These abilities were important in the establishment of different breeds. (See chart on page 93 for a detailed discussion of gaits.)

Mustang—Originally from the Spanish word *mesteño*, meaning stray animal. Some people use this term to refer only to wild horses, or horses removed from the wild, that show strong characteristics of the Spanish horse. Others use the word mustang to describe any wild horse, regardless of its characteristics. I use the terms mustang and wild horse interchangeably in this book, as most of the wild horse populations I discuss are descendants of Spanish horses, though other breeds have mixed in over time. Pejorative names for mustangs throughout U.S. history include broomtail and cayuse. Many mustangs are protected on federal lands.

Pony versus horse—Some breeds of smaller horses with particular characteristics are called ponies. However, small horses of other breeds under about 14.2 hands can also be referred to as ponies. Some people get sensitive about these distinctions, but whether a horse is a pony is often in the eye of the beholder. For example, the National Park Service manages the wild horses on the Maryland side of Assateague Island and calls them wild horses, while the Chincoteague Volunteer Fire Company manages the wild horses on the Virginia side of the island and calls them Chincoteague ponies. However, all the horses originated from the same herd. All modern horses fall under the species *E. caballus*, regardless of whether they are referred to as horses or ponies. I use the term pony loosely in the book, since wild horses are often under 14.2 hands and people have referred to them as both ponies and horses throughout history.

Nonnative versus native species—Most U.S. government agencies and much of the public consider the wild horse a nonnative species to the U.S. However, from a biological perspective, given that the modern horse evolved across the North American Great Plains in an ecosystem that is similar to the current ecosystem in that region, it follows that America's modern wild horse can be considered a native species or reintroduced native species. This classification controversy has policy implications for wild herds.

Wild versus feral animals—People are often quick to point out that America's modern wild horses are not actually wild but feral, meaning they are escaped domestic stock. As I will argue later, this view is shortsighted given that they are the same species as their wild ancestors and that most of today's population is wild born and has been for generations. The terms wild versus feral matter when it comes to management policy of modern herds. In this book, I use the term wild horses, not feral horses.

Burro—The Spanish word for donkey, which is a member of the genus *Equus* but a separate species from the horse. The scientific name for the domestic donkey is *Equus asinus*. Many wild burros are protected, along with wild horses, on federal lands.

Mule—The hybrid offspring of a male donkey and female horse. One in two hundred thousand female mules can get pregnant when bred to a horse or donkey, but male mules are infertile.[6] Mules were used for plantation work in the South.

PART I

EQUUS RISING
55 MYA TO 1500 CE[*]

[*]MYA: million years ago; CE: Common Era/AD

1

EVOLUTION OF THE HORSE

M ost kids would probably beg their parents to bring it home as a pet: a small mammal about as tall as a medium-sized dog that bounds on deer-like legs. Its back is hunched like a rabbit's and its feet are padded like a dog's, with three toes on its hind feet and four toes on its front feet. Its small eyes sit close together about halfway down its snout; its mouth is full of basic molars for nibbling; and it has a small, simple brain.

But there would be a major obstacle to the pet adoption process: this animal, the first horse, is long gone. Known as the dawn horse, with a hotly debated scientific name of either *Hyracotherium* or *Eohippus*, depending which scientist you ask, it emerged fifty-five million years ago—that's eleven million years after the extinction of the dinosaurs—and marks our story's beginning of how the horse shaped U.S. history. The incredible evolution of this rodent-like creature into the powerful modern horse sets the scene for this story in two important ways.

First, the horse's evolution itself is part of U.S. history in that it occurred primarily across the North American Great Plains.

Second, since the ancient horse evolved to live on the plains, it should be no surprise that the modern horse thrived upon reintroduction to the American West by Spanish *conquistadores*, both as a powerful tool humans used for exploration and conquest and, eventually, as a wild animal. There the modern horse found a similar environment, though a bit warmer and wetter, to the one that existed at the time its ancestors last set foot on the continent thousands of years earlier.[7] Understanding the horse's adaptations to this environment helps us to make sense of how the animal could have had such a tremendous influence on the establishment and growth of the U.S., as well as on American culture.

When the dawn horse emerged, the climate of what is now the western U.S. was drastically different from the dry, open landscape we know today. Think tropics. It was hot, humid, wet, muddy, and covered in jungle-like trees and plants, resembling the jungles of today's South America more than today's Wyoming.[8]

The dawn horse was a browser, meaning it used its simple, short teeth to crush and chew succulent leaves and soft berries. Its widespread, padded feet kept it from sinking in the muddy swamps and getting stuck, just as snowshoes prevent a person from sinking into snow. It did

JULIA SOPLOP

14

not need to run long distances or have excellent vision to escape predators, because it could just bound under the thick foliage to hide if danger arose. We know this animal was well suited to its environment, because fossil records show it reproduced and spread all over the place.

But eventually the global climate began to change. Over the course of millions of years, Earth grew cooler and drier. These changes affected the horse in numerous ways. Luckily, it had an incredible ability to adapt to new circumstances.

The horse's extensive genetic variation resulted in a huge range of physical traits. At one point in time, at least twenty horse species existed in North America alone, many of which differed considerably from one another and occupied different ecological niches. Even within a species, horses could vary quite a bit. One fossil bed in Nebraska revealed two horses of the same species with a striking difference— one had one toe on each foot, while the other had three toes on each foot.[9]

The wide range of physical characteristics found among horses meant that whenever the environment changed, at least some variations of the horse continued to tolerate and thrive under the new conditions. The horses with traits that allowed them to survive in new environments left behind offspring with those traits. The animals with traits that were not helpful for survival in new environments died before producing many offspring, and eventually those traits disappeared from the gene pool. This process is called evolution by natural selection.

As the climate continued to change, becoming cooler and drier, grass eventually overtook the leaves-and-fruit-filled woodlands; horses began to lose their food sources. Some horses died out. But others adapted to eating grass, thanks to two interesting characteristics that emerged in their skeletal and digestive systems.

Grass presents a significant problem.[10] Like most plants and animals, it has a built-in defense mechanism to try to keep from getting eaten. It extracts silica (sand) from the soil to make itself abrasive, like sandpaper. It acts like sandpaper, too, eventually wearing teeth down to nothing. When animals' teeth wear down, they lose the ability to eat. They die.

Many animals that depended on the woods for food were not able to adapt to eating grass, including some types of horses. They died out. But the horses that did adapt to become grass eaters, called grazers, had larger, more complex teeth with ridges and grooves that allowed them to grind their food, like modern horses. These teeth took a long time to wear down, too, even against their new sandpaper-like food source.

Over time, horses' teeth evolved to be even longer. They also began to function in a specialized way: as grass wore the teeth down, they would continue to push up through the gums to maintain a grinding surface for eating. It would take decades for grass to wear away these large, intricate teeth to uselessness. Longer teeth meant a longer potential

lifespan.

But it takes more than good teeth to live off grass. Around the same time the teeth adaptations developed, the cecum, or hindgut, also emerged as part of the digestive system. A large cecum allows horses to absorb more nutrients from nutrient-poor foods than other grazers, meaning they can live on foods that are not nutrient-rich enough for other animals, such as modern cows, to subsist on.[11] This characteristic allows horses to live in harsh, desolate areas where other animals may not survive, but they must eat constantly to take in enough nutrients.

Horses born with this combination of teeth and digestive system changes had an advantage over other horses, because they could eat the foods available to them in their new, grass-filled environment.

Escaping predators is as essential to survival as eating. The shifting climate changed both the horse's predators and its ability to evade them. Saber-toothed cats and dire wolves were large and fast, and horses could no longer bound under thick foliage to hide.[12] Much of the foliage was gone, replaced by grass. Over time, the horses that survived these new threats emerged with numerous additional adaptations that helped them escape the jaws of predators.

The eyes moved up and outward in the skull, allowing the horse to watch for predators while its head was down to graze or drink.[13] The eyes also increased in size and complexity. The horse became sensitive to quick movements and developed a wide range of vision.

The little dawn horse of fifty-five million years ago had padded feet with numerous toes on each one, which helped it to stay aloft and succeed in a muddy environment. But as Earth cooled and became more arid, the landscape slowly transformed from mud to the dry, hard ground of today's plains. Over time, the horse adapted by reducing its toes down to one, meaning horses whose genes had mutated to produce one toe instead of three were better suited to survive the new environment than their multi-toed counterparts, which eventually died out. This toe—a hoof— was specialized for speed on the hard ground, helping the horse to escape new predators. (Having one toe classifies the horse as an odd-toed ungulate, an unusual distinction it shares with the modern rhinoceros and tapir.)

Other speed innovations included a straighter back and longer legs. Several leg bones also fused together, so the legs could not rotate as easily as a human's but became extremely efficient for running.[14]

The horse's body size also increased over time. Larger horses were better able to defend themselves against increasingly larger predators and, according to Bergmann's Rule, may have been able to better maintain their body heat in a colder climate. Although some small horses existed alongside larger ones for quite a while, only larger horses ultimately survived these new environmental and ecological challenges.

Eventually the horse developed a much bigger, more complex brain in comparison to the dawn horse's. These adaptations may have been important for making rapid decisions in response to danger in the exposed grasslands and for the elaborate social relationships horses have in their bands, the groups of family and allies they live with.[15] Fortuitously, the horse's brain adaptations that allowed it to navigate social relationships within its own species would contribute to its ability to form relationships with humans later.

By the time *Equus*, the genus in which our modern horse belongs, emerged three to five million years ago, and *Equus caballus*, the modern horse species that includes both domestic and wild horses, emerged approximately 1.7 million years ago,[16] it was packed with all of these evolutionary adaptions for success in its drier, cooler climate, and eventually our modern world.

Some species of *Equus* wandered back and forth across the Bering Land Bridge into Asia, Europe, and Africa starting about two million years ago and began to differentiate, eventually resulting in our modern zebras, wild asses, the onager, and Przewalski's horse.[17]

Equus survived in North America until around the end of the Ice Age. But, mysteriously, it disappeared from the continent between eleven thousand and eight thousand years ago.[18] This disappearance occurred alongside the extinction of nearly one-fourth of large-bodied mammals in North America, an event known as the Quaternary Extinction.[19] A great controversy surrounds the reason for the horse's disappearance from the continent, but many scientists agree that a combination of abrupt climate changes at the end of the Ice Age and hunting by the humans, who arrived at least fifteen thousand years ago and possibly as early as forty thousand years ago,[20] did them in.

Despite the fact that early peoples in North America lived in small, scattered groups, evidence shows they developed and spread an effective type of stone weaponry

SIGNIFICANT ADAPTATIONS OF EQUUS		
	Dawn Horse	Equus
Teeth	Basic molars for nibbling berries	Large, complex teeth to grind grass
Digestive system	No cecum	Large cecum to digest nutrient-poor foods
Eyes	Small eyes; halfway down snout	Big, well-positioned eyes to spot danger
Toes	Three toes on hind feet and four toes on front to stay aloft in mud	One toe (hoof) to run on hard ground
Legs	Small, deer-like	Long for running
Body size	Similar to a medium-sized dog	Large for safety and to maintain body heat
Back	Rounded to bound like rabbit	Straight for running
Brain	Small and simple	Large and complex to navigate threats and social relationships

used to hunt megafauna, known as Clovis points, across the entire continent in a span of just two hundred years.[21] One author posited that even without the dramatic climate shifts that occurred, Clovis hunters alone could have brought the megafauna to extinction within a few centuries.[22] Thankfully, before going extinct in North America, ancestors of the modern horse walked across the Bering Land Bridge and spread through Asia and Europe.

The last known species of ancient *Equus* to go extinct in North America was *E. lambei*, known as the Yukon horse for the location it was found, about eight thousand years ago.[23] Scientists first classified *E. lambei* as a unique species based on the visible physical characteristics of its fossils. More recently, however, molecular biologists have found, using mitochondrial DNA, that *E. lambei* is genetically equivalent to *E. caballus*, the modern horse.[24] Although *E. lambei* looked more like Przewalski's horse of Mongolia than a modern Quarter Horse, their genes indicate *E. lambei* and *E. caballus* are one in the same.

So the last horse species to live in North America, which humans were likely in part responsible for finishing off, could be considered the same species the Spanish later reintroduced to the continent. Whether our modern wild horse should be considered a native, wild species—one that evolved in the U.S.—or a nonnative, feral species—an escaped domestic animal exotic to the U.S.—is controversial to this day and has implications for wild horse management policies, as we'll discuss later.

2

DOMESTICATION AND COLORATION

More than five thousand years ago, humans across Europe and Asia began to catch and domesticate horses from local wild bands of *E. caballus*, which varied considerably in size and appearance across different regions.[25] Over the centuries, European, Asian, and African peoples, once humans brought the modern horse to the African continent, selectively bred horses to have characteristics valued for their particular uses.

Prior to domestication, there was little color variation among wild horses. A combination of archeological evidence, in the form of ancient cave paintings, and genetic evidence, in the form of fossil DNA testing, has revealed that the wild type—the most common base coat coloration in wild horses before domestication—was bay.[26] Bay is a brown horse with black on what are known as the points: the lower legs, edges of ears, mane, and tail. Black

23

coloration, as well as leopard spotting, as seen on some modern horses, also existed in the wild prior to domestication.[27]

Wild horse in Montana's Bighorn Canyon National Recreation Area with dun coloration that includes primitive striping on the withers and a dorsal stripe down the middle of the back.

Both modern wild equid coloration and prehistoric cave paintings suggest early wild horses were also likely to be dun, meaning they had a gene mutation that lightened their body color to yellowish from dark brown or grayish from black, added striping on the legs and withers (shoulders) and down the middle of the back, and sometimes contributed spider webbing to the face.[28] However, scientists have not yet isolated how to test for the dun gene in prehistoric fossils.

Horse coat coloration exploded shortly after domestication.[29] Evidence indicates that humans valued novel colors in horses. When domestic horses were born with mutations that gave them non-wild-type coloration—colors and patterns that in the wild were detrimental to survival and naturally selected out—humans quickly bred them, adding the unusual colorations to their domestic stock. The more that humans bred their domestic horses that had less-common colorations, the more mutations related to the genes that coded these colorations emerged in their offspring, resulting in even more color variations in subsequent generations.

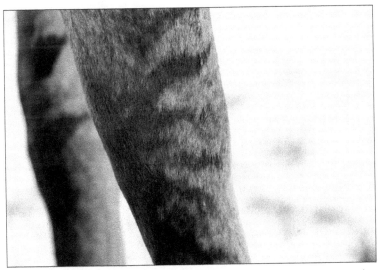

Primitive zebra-like leg striping of a domestic Quarter Horse on the Wyoming ranch of Steve and Nancy Cerroni.

Aside from color, people in different regions changed modern horses in a variety of other ways. European horses were bred to carry heavily armored knights into battle during the Middle Ages, around 500-1500 CE. These horses were large, slow, relatively calm, and ate particular grains. We call them draft horses or cold bloods. (The term cold blood does not suggest a horse is a cold-blooded reptile; it describes a horse's calm temperament. Hot blood describes a higher-strung horse.)

The desert horses—hot bloods—that emerged in the Middle East (Arabians) and North Africa (Barbs, which were originally offshoots of the Arabian) were small, light, fast, and temperamental. Like their ancient ancestors that evolved across the plains of North America, they possessed the strength and stamina required to survive and thrive in harsh, arid conditions.

Around 700 CE, Moors from North Africa, mounted on Arabians and Barbs, invaded and conquered Spain. The Spanish recognized the advantages of the Moors' tough, quick horses over their own larger ones and began to breed them together to produce a new kind of horse—a warm blood. By combining a bit of their draft horse blood with Arabians and Barbs, they created the Jennet. It was smaller and more compact than their large horse—around 13-14 hands—and became known over the next eight hundred years for its exceptional intellect, stamina, and speed.[30] The Jennet is sometimes referred to as an Iberian horse or Andalusian. We'll refer to it as the Spanish horse.

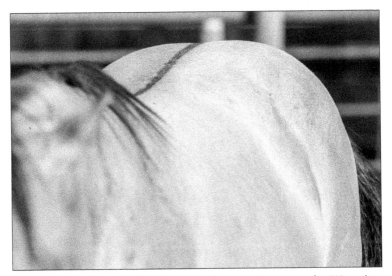

Primitive dorsal stripe of a domestic Quarter Horse on the Wyoming ranch of Steve and Nancy Cerroni.

The Spanish horse was strong and stocky, with a flat forehead and widely separated eyes.[31] It had small ears; front legs that came together in a V-shape instead of a square; a tapered muzzle, sometimes concave like an Arabian's, sometimes convex like a Barb's; and a short back due to one fewer vertebra than most other breeds (technically two vertebrae fused together), thanks to its Arabian heritage. They were fast, hardy, and had tough hoofs that did not require horseshoes. They were bred in all colors.

The Spanish horse, bred mostly in Andalusia, became the original stock for Spanish exploration of the Americas.

PART II

THE WEST
1500s-1700s

3

RETURN OF THE HORSE

T raditional history books will never let us forget that in 1492, the controversial Christopher Columbus sailed the ocean blue. But it was in 1493 that he brought the first Spanish horses across the sea.

Only about one-third of the horses shipped across the Atlantic during those early years of conquest survived the passage.[32] The survivors made for especially hardy stock, and ranches sprang up all over the West Indies, churning out Spanish horses for the "New World."

The horses had maintained many of the characteristics of their North American ancestors, including incredible stamina and the ability to survive solely on nutrient-poor grasses and little water—traits that gave the Spanish *conquistadores* an advantage in exploring dry, desolate areas of the Americas that other breeds, such as the large cold bloods of the French or English that required grain supplements and horseshoes, could not have offered.[33] The same features also made the Spanish horses extremely successful and self-sufficient when some eventually escaped and reverted to the wild in a landscape similar to the one in which their ancestors had evolved.

The fact that native populations had long ago claimed these lands meant nothing to the Spanish, who viewed the "New World" as theirs for the taking. They used their horses to search for gold and seize territory for the empire. As the *conquistadores* and other European explorers made their way through South and Central America, they conquered most of the native peoples with relative ease, using the horse, time and again, as their secret weapon.

In 1519, eleven stallions, five mares, and one colt (born on the ship) became the first modern horses to set foot on the North American mainland, and the continent would never be the same.[34] The horses arrived in Vera Cruz, Mexico, with Hernán Cortés about seventy-five hundred years after the last of their ancient relatives died out in North America. Although Cortés arrived in Mexico with just five hundred men and sixteen horses (the colt escaped), he conquered the Aztecs within two years, giving credit to his horses, calling them: "our salvation," and supposedly saying, "Next to God, we owed our victory to our horses."

Not only were his horses hardy enough to carry their riders through the arid landscape, but they played an even more critical role than perhaps expected in the conquest: they terrified the Aztecs into submission. Since modern horses did not exist in the Americas at that time, the Aztecs had no idea what animal was charging at them, believing a horse and rider were a monster that could split into two.[35] This fear factor meant everything. A chronicler of the expedition, Bernal Díaz del Castillo, wrote that the horses were "our fortress" and "our only hope of survival."[36]

In Cuba, Pánfilo de Narváez fought off thousands of native warriors by galloping through their forces ringing bells.[37] He went on to conquer the island. In Peru, Francisco Pizarro used a similar technique to overcome an Incan army of eighty thousand with just sixty-two men on horses and 106 on foot.

The indigenous nations that fell quickly were largely agricultural societies organized around central leadership.[38] Overtake the leadership; overtake the people underneath. The Europeans simply slaughtered those who refused to assimilate and fall into servitude.

In fact, Europeans wiped out so many people so fast across the Americas, through a combination of violence and disease, that they changed the global climate in the process.[39] Within about a century of setting foot on American soil, Europeans and the diseases they brought along with them depopulated the Americas of indigenous peoples by up to 90 percent, killing off an estimated fifty-five million people—about a tenth of the global population at that time. The land those peoples farmed was suddenly abandoned, and it immediately reforested. The new forests rapidly began to suck carbon dioxide out of the atmosphere, contributing to a decrease in global temperatures, which ushered in the Little Ice Age of the 1600s.

When the Spanish trailed up into what is now the U.S., they probably thought they would use their horses to conquer the native peoples of the North as swiftly as they had in the South. Fear factor aside, in the sweeping, dry landscapes of the American Southwest and Great Plains, where their ancient ancestors had evolved, Spanish horses gave incredible power to those who could master them.[40]

In 1540, Francisco Coronado tracked up through New Mexico on an exploratory expedition with fifteen hundred horses. The Native Americans the group encountered were so overcome by the power of the horse that they spread its sweat onto their own bodies, hoping to acquire its strength.

In 1598, Juan de Oñate caravanned northward through Mexico with several hundred horses to start a settlement outside of what is now Santa Fe, New Mexico. The settlers quickly enslaved the Pueblo, forcing them to work the ranches they were establishing. By this time, the Spanish understood the full power of the horse in their control over native peoples, so the Spanish government immediately banned Native Americans from riding. But by 1621, the government changed its tune after failing to convert many Pueblo to Catholicism, offering them the right to ride in exchange for conversion.

The Spanish made a grave miscalculation here in their quest to rule the continent. Despite recognizing that the horse gave them the ability to dominate the Pueblo, the Spanish, in addition to allowing converts to ride, also insisted the Pueblo learn to care for and breed the ranch horses.[41] Every once in a while, a disgruntled Pueblo would gallop off into the hills on one of the ranch horses, sharing his horsemanship knowledge with others. By the 1640s, the Pueblo had shared their expertise with nearby indigenous peoples and taught some Navajo and Apache chiefs to ride.[42] By the 1650s, Apaches were stealing horses from the Spanish settlements and other tribes. By 1676, the Spanish had lost so many horses to raids they barely had any left in New Mexico.[43] Seventy years after their arrival in the region on horseback, the Spanish were losing control of their most powerful weapon.

Perhaps the most significant moment for the horse in the

West, however, arrived during the Pueblo Revolt of 1680. The Pueblo and their neighboring native peoples rose up against their Spanish enslavers, killed four hundred of them, including twenty-one priests, and chased the rest of the two thousand Spanish settlers out of the area. The Spanish fled Santa Fe so fast, they left behind around three thousand horses.

In the chaos, the Native American tribes, who remained free from Spanish incursion for more than a decade after the revolt, gathered up many of the discarded horses for themselves, perhaps unaware of how fully this event would change their lives. Numerous horses also escaped to the wild, becoming the foundation of North America's modern wild horses.

Wild horses roaming free in Montana.

Known as the Great Horse Dispersal, this moment would transform the history of the Southwest, the Great Plains, and the rest of what would become the U.S. The Spanish stayed out of the region for more than a decade. When they returned twelve years later, the Horse Nations were waiting for them.[44] Ironically, the source of Spanish power in the Americas would become their downfall.

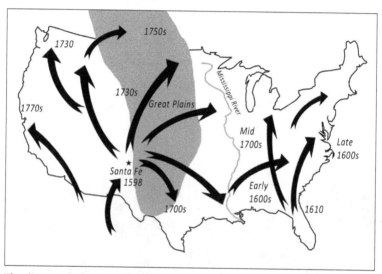

The dispersal of Spanish horses throughout North America (approximate dates).

4

RISE OF THE HORSE NATIONS

After the Pueblo Revolt of 1680, horses spread like wildfire through the Native American nations of the Great Plains. By 1700, all the tribes in Texas had horses.[45] By 1750, tribes as far north as the Canadian plains and as far northwest as modern-day Oregon and Washington had acquired them. The age of the horse was in full swing across the North American plains.

While some native peoples of the plains used the horse for basic purposes of food or transportation, others embraced the full power of the animal to transform their cultures. These peoples—the Apache, Blackfoot, Cheyenne, Comanche, Crow, Kiowa, Nez Perce, and Sioux, among others—became known as the Horse Nations. As masters of the horse, they would rule the plains for nearly two centuries and become forever linked to the spread of America's modern wild horses.

Before the reintroduction of the modern horse in the

West, most of the plains nations were semi-nomadic, meaning they had permanent dwellings but spent part of the year on the move. Many lived in wooden structures at the edges of the grasslands, near rivers or forests, and made treks into the plains to hunt buffalo, which they depended on for many aspects of life, including for food, hides to make clothing and tipis, and dried buffalo chips to fuel their fires. (The animal colloquially referred to as the buffalo is actually the American bison. Although the term buffalo is scientifically inaccurate, it has become a commonly accepted term for the American bison. To avoid confusion, given its common usage in this context, we will refer to the animal as a buffalo.) Buffalo hunters would pack small tipis and supplies in travois, triangular frames constructed from lashing wooden poles together, pulled by dogs.

The Great Plains, however, were too harsh for year-round living—not enough water or wood, too much wind and drifting snow—and the indigenous peoples did not have methods to transport entire villages long distances.

The horse changed everything for these tribes. First, it opened up the vast plains. Instead of being forced to live on the outskirts and venture into the plains for short hunting trips, the Horse Nations were suddenly able to make full use of the landscape. Horses could pull much larger, more heavily packed travois than dogs. They acted not only as transportation for hunters, but also as vehicles to move entire tribes and their belongings, covering long distances to the next watering hole. Whereas dogs required precious meat, horses could subsist on grass. They allowed tribes to migrate with the buffalo year-round, and many of the Horse Nations became entirely nomadic.[46]

Horses also transformed the buffalo hunt itself. Hunting on horseback was much more efficient than hunting on foot, and tribes began to acquire and train horses specifically for the hunt.

Although many depictions of the Horse Nations show men riding horses, women became expert riders, too. Their skills extended far beyond just moving their families and belongings from camp to camp on horseback.[47] Native American women were known to gallop across the plains roping buffalo, antelope, and wild horses from the saddle.

Once the nations recognized the horses' incredible value, they treated them as a form of currency. For the first time, these tribes could become wealthy. They traded horses for goods with other Native Americans, as well as white explores and settlers they encountered.

The horse eventually became an indicator of social status. Horse stealing was considered an art, one that brought great honor to a talented horse thief. The more horses a warrior collected—through long-strategized plans, middle-of-the-night-raids, and sneaking into tipis to whisk away staked horses—the more respect he earned.[48] Guarding one's own horses from rustlers also became a full-time occupation.[49]

Some of the Horse Nations believed the most sacred horse was the Medicine Hat stallion, also known as a War

Bonnet—a mostly white horse with a patch of color at the top of its head and over the ears like a bonnet, and sometimes a shield of color splayed across its chest. Medicine Hats often had blue eyes, as well. They were said to have mystical qualities that made a warrior invincible, thereby making them particular targets of thievery.

The way the indigenous peoples of the plains handled their horses contributed to the swift rise of wild horses. There were no fences on the plains, and the Horse Nations would let their animals roam freely to find food and water, then recapture them when needed. Horses often ran away, starting or joining wild bands. The Horse Nations also lost many horses to the wild during the relentless process of stealing them from each other, settlers, and anyone who came looking to subdue the native peoples of the plains.

When the Horse Nations wanted to increase their herds beyond what they stole or bred—they preferred trained horses when available—they captured wild horses. When individuals or entire indigenous nations perished from diseases brought by Europeans, such as smallpox, their herds disappeared into the wild.[50]

There were no modern wild horses in the West before the native peoples gained access to them from the Spanish. The Horse Nations' practices of raising, stealing, and trading horses triggered the rapid dispersal of wild horses across the West. As horses proliferated, they reshaped the Great Plains, both by opening them up to human use and shifting power to those who most fully harnessed their strength.[51]

5

RISE OF AMERICA'S MODERN WILD HORSES

V isitors to the Great Plains during the 1700s and 1800s
were floored by the sheer number of wild horses they
encountered. Some reported seas of horses in constant
movement, clear to the horizon.[52]

Spanish settlers initially called the wild horse *mesteño*,
meaning stray animal. Eventually, English speakers
adapted this word into the one we still use today: mustang.
The original mustangs that scattered throughout the West
were small, scrappy, intelligent Spanish horses. These
animals—the products of millions of years of evolution
across the plains of North America, further years as desert
horses in Africa and the Middle East, and hundreds of years
of breeding in Spain—were prepared to live and propagate
in the western landscape they found when they escaped
their owners and became wild.

Mustangs spread and multiplied with such success
throughout the West that at their peak in the early 1800s,

they numbered at least two million, with about one million in Texas alone.[53] Some historians estimate population numbers two or three times that high. In fact, some explorers simply marked early maps of what is now Texas: "Vast Herds of Wild Horses" or "Wild Horses."[54] One area of Texas was referred to as "The Mustang Desert."[55]

Were these horses native or nonnative to North America? It depends whom you ask. Many people are quick to label them as nonnative and feral—a precarious status, as you'll read later. In the U.S., we tend to consider a plant or animal native if Europeans observed it when they first arrived on the continent.[56]

This concept has largely shaped government policies on wildlife management over the years. For example, a 1963 policy memo, commonly referred to as *The Leopold Report*, recommended that as a primary goal of the National Park Service (NPS), "the biotic associations within each park be maintained, or where necessary recreated, as nearly as possible in the condition that prevailed when the area was first visited by the white man. A national park should represent a vignette of primitive America."[57] This report has had profound influence over NPS policies to this day.

Since there is no evidence that wild horses were living on the continent at the time the Spanish arrived, they do not fit the bill for a species worth preserving, according to *The Leopold Report*. Labeling the offspring of escaped Spanish horses an unnatural or a nonnative part of the ecosystem might seem simple enough using this archaic definition of

A blue roan band stallion covered in scars from a lifetime of fighting with other stallions. Roan horses have white hairs mixed into their base coat color. Hair grows back in the base color after an injury, in this case black.

what belongs here.

But an accurate answer is considerably more complicated and controversial. *E. caballus* evolved across the North American plains. The last ancient horse to roam the continent died off eight thousand years ago, likely in part due to human interference. Through domestication, humans altered the appearance of *E. caballus*, but the horses the Spanish brought to the continent were genetically equivalent to the last horse species to live in North America.

From a biologist's perspective, the determination of whether a species is native has nothing to do with whether white men laid eyes on it when they first arrived somewhere. It depends on where the species originated and whether it co-evolved with its habitat.[58] It is difficult to argue that the horse the Spanish reintroduced after just a blip on the evolutionary timescale is a nonnative species from a biological standpoint. Evidence places the horse's evolution in North America; demonstrates that the environment in which *E. caballus* emerged does not differ much from the environment of the West today; and suggests humans contributed to the horse's extinction from the continent. [59]

Whether modern wild horses should be considered feral or wild is a slightly different question but with similar policy implications. The term feral is used to describe a domestic animal that has escaped and become wild. Feral animals are managed differently than truly wild animals. While many U.S. government agencies and much of the public classify wild horses as feral, this definition is simplistic.

The first modern wild horses in the U.S., and many subsequent wild horses throughout the centuries, were escaped domestic horses. Perhaps one could argue those individuals should be considered feral. But what about their offspring, animals born in the wild in the place their ancestors evolved and in a similar ecosystem? And what about the subsequent offspring of those animals that have now been co-evolving with their habitats for hundreds of years? Can these animals be labeled feral, even if they are the same species as the unquestionably wild horses that evolved here? Suddenly, the concept of feral seems contrived and meaningless, or at least ill-fitting for America's wild horses.

Perhaps the most appropriate terminology to describe today's wild horse in the U.S. is a reintroduced native species. While it may seem unnecessary to debate such distinctions and labels, these terms will have policy implications down the line in the determination of whether wild horses should have the right to exist on public lands, who should manage them, and how we should treat them.

We do not know whether early witnesses to the West, a region packed with reintroduced horses, cared about or questioned the horses' origins. But we do know what they saw were horses that had quickly reverted to wild horse behavior, a common phenomenon now supported by research.[60]

Wild horses live in family bands with complex social dynamics. Bands are made up of about three to ten individuals, including a stallion, at least one mare, and their young offspring. There is still plenty of literature today referring to a band as a harem, controlled by one dominant stallion that fathers all the offspring. While casual observations of aggressive stallions in the wild may propagate this idea, the characterization is antiquated; social dynamics within a band are much more nuanced.

For example, one study on Assateague Island, which is split between Maryland and Virginia, revealed that band stallions examined were the biological parents of just half the offspring in their bands.[61] Mares often snuck off to breed with stallions other than their band stallions.

A dun mare of Montana's Pryor Mountains, with leg striping, watches over two sleeping foals, one of which is not her own.

Mares also have their own bonds and alliances within their bands, as well as their own pecking order. They help care for foals that are not their own, too. Mares often subtly govern the movements of their bands, while stallions serve as the body guards, picking up the rear.[62]

When colts grow into mature stallions, their fathers run them off. Some bachelors live alone. Others form bachelor bands. Eventually they begin to challenge band stallions for mares. At first their efforts might be pitiful, but as they gain strength and fighting experience, they can become formidable foes against aging band stallions. Stallions

47

spend their lives fighting to win over, breed, maintain, and reclaim mares.

The brain development that occurred throughout the horse's evolution enables it to foster these complex social relationships. Many scholars have pointed out that the capacity of horses to form social bonds with each other also enables horses to form strong bonds with humans—a key characteristic that, since domestication, has elevated the partnership between horses and humans to a level above other animal-human relationships.

A band stallion chases off a bachelor stallion encroaching on his mares and foals.

The social nature of the horse made possible the dance between wild horses and the Horse Nations, each influencing and giving power to the other as the rest of the continent came under European control. While the horse had varying influences on different native peoples, we'll take a closer look at the way it influenced two in particular: the Comanche and the Nez Perce.

6

THE COMANCHE

A round 1680, the arrival of the horse transformed the Comanche perhaps more fully than any other Native American nation. They quickly became some of the most skilled horsemen—arguably *the* most skilled horsemen—in the world, and used their abilities to overtake the southern Great Plains.

Prior to acquiring the horse, Comanche culture had not changed much in the thousands of years since their ancestors crossed the Bering Land Bridge into North America.[63] If they had been strong warriors, they would likely have fought for a more hospitable place to call home than the Wind River area of modern-day Wyoming.

They were hunter-gatherers who lived off what they could find, including rodents. They also hunted buffalo on foot by lighting fire to the grasslands and chasing them over bluffs. These inefficient hunting techniques meant they had little time to spend developing a cultural life, such as making pottery, weaving baskets, or assembling intricate clothing.

They never planted crops. They used the dog travois. They did not have any central leadership over the separate bands that totaled around five thousand people.

They eked out a living and survived.

And then the horse arrived. The Comanche acclimated to the horse so rapidly and intensely that their culture began to revolve almost exclusively around its use.[64]

They became expert buffalo hunters on horseback. No more setting fire to prairies. No more chasing buffalo on foot, hoping a few might end up at the bottom of a gorge. Now they could train a fast horse to gallop beside a buffalo and stay with that specific animal while a hunter dropped the reins and gored it with a fourteen-foot lance or shot it with arrows.[65]

Their culture depended heavily on buffalo for food, shelter, and clothing. Becoming efficient hunters meant the Comanche no longer had to devote all their time to hunting. Now they also had something worth fighting for: the plains that held as many as sixty million buffalo.[66]

The Comanche borrowed their basic riding culture from the Spanish but took their horsemanship skills to a significantly higher level. Thanks to the horse, they soon transformed into talented mounted warriors, and by the late 1600s began to push out other nations to take control of the prized southern plains.[67]

By 1750, the Comanche had established a territory that

extended through much of Texas and portions of New Mexico, Colorado, Kansas, and Oklahoma.[68] They expertly and violently patrolled and guarded this territory, known as Comancheria, against encroachment from neighboring native peoples, as well as Spanish and other European settlers.

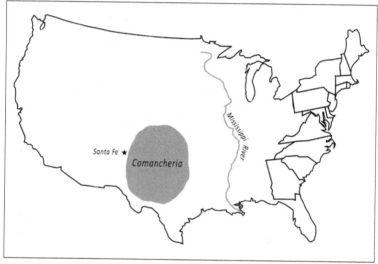

Approximate boundaries of Comancheria in 1750. (Original thirteen British colonies are outlined).

They started their children riding young. By four or five, Comanche kids had their own horses.[69] Boys immediately began training to master a variety of skills and tricks to prepare them to be warriors, such as learning to pick up fallen fellow warriors off the ground at full tilt with the sweep of an arm and ride them to safety.

One of the most impressive skills they developed was to

slide over and hang off one side of the horse by their feet, using the horse as a shield while shooting arrows at an enemy from under the animal's neck—all at full gallop.[70]

Comanche women were excellent riders and owned their own horses, as well.[71]

No one in the region could match this level of skill: not the settlers; not the military; not most other native peoples except perhaps the Comanche allies, the Kiowa. For starters, white settlers were not even attempting to live out in the open, treeless, waterless plains the way the Comanche had learned to do with their horses. And anyone who attempted to encroach on Comancheria, with their rudimentary Kentucky rifles that required dismounting to discharge, found themselves sorely overpowered by the Comanche, who could unleash twenty arrows on horseback in the time it took to shoot and reload a rifle.

The Comanche understood horses meant power and were unwilling to let others wield that power over them. To increase their own supply and deplete the Europeans and other Native Americans of theirs, the Comanche became expert horse thieves. A Comanche warrior might acquire a hundred or two hundred horses. A chief might acquire fifteen hundred.[72]

Comancheria pressed up against Spanish settlements of the Southwest, and the Comanche regularly raided them for horses. Sometimes they would simply scare off entire horse herds, known as stampeding, just to relieve their foes of them. Many of these horses would become wild. Sometimes the Comanche went on specific missions to steal prized war or buffalo horses to add to their own herds.

They would also take white captives during these raids, either to enslave or later sell for ransom. But no one could compete with these horsemen to retaliate the raids. Parties that tried to search down the Comanche across the plains for retribution for horse stealing or kidnapping often found themselves relieved of their own horses and left to die of thirst or starvation in the middle of the grasslands.[73] Hardy horses meant the only chance of survival on the parched, endless plains.

Time and again, the Comanche depleted the Spanish settlements of horses to the point that the Spanish had to order thousands from breeding ranches in Mexico to replenish their stocks. When the Spanish became desperate, the Comanche would sell the Spanish back their own horses.[74] The Comanche bragged they only allowed the Spanish to continue living nearby so they could serve as their horse breeders.

For decades, the Comanche made annual trips into Mexico to raid ranches for even more horses, trailing them back to Comancheria to increase their herds and use for trading.

They would also capture wild horses and were known to be able to break a mustang in a couple of hours, though not gently.[75] They would often wait until a wild horse was weakened after a winter of little food, then chase it down, rope it, and choke it to the ground until the animal stopped

struggling. Then they would slacken the lasso around its neck, let the animal stand, slowly approach it, stroke its face, and breathe into its nostrils. They would then slip a simple bridle around the lower jaw, climb on, and take a ride.

The Comanche had such a tight hold over Comancheria that they relentlessly beat back the Spanish, eventually forcing them out of the territory. They managed to keep everyone else, including the French, English, and eventually Americans, from meaningful settlement of the southern Great Plains until the late 1800s. Although the U.S. had already acquired the land west of the Mississippi in name by that time, the government soon understood it would never actually control that land unless it could subdue the Horse Nations. In the Southwest, doing so meant trying to figure out how to remove the Comanche from the position of power they had earned through their incredible use of the horse. Comanche territory was the last region of the country the U.S. military would bring under its control.[76]

7

THE NEZ PERCE AND THEIR APPALOOSAS

B y 1730, while the Comanche were aggressively carving out Comancheria across the southern plains, the Nez Perce had acquired horses in the Northwest. (Although they never pierced their noses, French trappers who encountered them mistakenly called them Nez Perce, meaning pierced nose, after confusing them with other Native Americans.)

The four thousand people of the Nez Perce lived in small bands scattered across the Columbia Plateau of modern-day Washington, Oregon, Montana, and Idaho.[77] Hemmed in by the Bitterroot Mountains of the Rockies to the east and the Cascade Range to the west, the plateau is a high shelf of flood basalt formed by lava that poured out of the earth for millions of years. Its dramatic landscapes vary from steep gorges cut from the Columbia River to riverside forests to temperate mountain meadows, rich with grass.

The Nez Perce were hunter-gatherer-fishermen, who lived primarily on camas bulbs, berries, and salmon.[78] The plateau offered plentiful space and natural resources for them to prosper. They did not have a particularly warring culture and, aside from a few skirmishes with aggressive neighboring nations, lived a relatively peaceful existence in their isolated environment.[79]

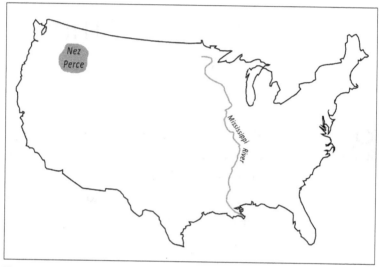

Nez Perce ancestral lands on the Columbia Plateau of modern-day Washington, Oregon, Montana, and Idaho.

The horse transformed their lives somewhat differently than it did the lives of the Comanche and other Horse Nations.

The grasslands of the plateau provided excellent grazing ground, allowing the Nez Perce to raise large numbers of horses without much need to rustle them from their neighbors. Though raiding did sometimes occur, the natural barriers of the Bitterroots and Cascades also protected the Nez Perce from being the victims of the constant horse stealing happening across the plains.[80]

History would prove the Nez Perce culture to be innovative and adaptive.[81] Unlike the other Horse Nations, who were more apt to collect horses than purposefully breed them, the Nez Perce began a selective breeding program.[82] They developed their own tools and procedures to geld, or make infertile, inferior stallions—methods white explorers would later praise as superior to their own.[83] They bred their horses to become sleek, strong runners, that were versatile for travel, hunting, and battle. At 14-15.2 hands, these horses, eventually called Appaloosas after the nearby Palouse River, were larger than most of the Spanish horses of the plains.[84]

While some Appaloosas were solid colors, many exhibited various forms of spotting, a characteristic for which they became known. There is a common misconception that the Nez Perce actually developed the spotting of the Appaloosa through selective breeding, which is not the case. As we established earlier, genetic testing has demonstrated that leopard spotting is a primitive horse coat coloration.[85] Cave paintings in France dating back twenty thousand years also depict spotted horses, as does art in Greece and France from 1400 BCE, Italy and Austria from 800 BCE, China from 500 CE, and Persia from 1300 CE.[86] This spotting was a color variation of the Spanish horse.

The Nez Perce believed their spotted horses made the

APPALOOSA COAT PATTERNS	
Leopard	White with dark spots over the entire body
Spotted Blanket	Dark body with white "blanket" over hips and dark spots on the blanket
White Blanket	Dark body with white "blanket" over hips
Frost	Dark body with white hairs mixed in over hips and sometimes over other parts of the body; sometimes including spots on these areas
Snowflake	White spots on dark body
Marble	White hairs mixed into darker body color; often lighter on back and hips and spotted; often darker on edges of body

best buffalo hunters and warhorses and were therefore the most valuable. So they purposely bred for spots. Appaloosas are distinguished by their distinctive coat patterns: leopard, spotted blanket, white blanket, frost, snowflake, and marble.

The Nez Perce became extraordinary mounted hunters, warriors, and all-around horsemen. Like the Comanche, their training began at a young age. Children were strapped to saddles to sleep during long family travels by horseback.[87] Like the Comanche, they also learned early to slide off to one side of their horses and shoot arrows from under their horses' necks. In one fluid motion, warriors could also hop off their horses, fire single shots from their rifles from the ground, and remount. The Nez Perce enjoyed a wild pastime of racing and betting on their horses, too.[88]

The horse increased the mobility and range of the Nez Perce. Although the Columbia Plateau was not buffalo territory, the Nez Perce began to ride their Appaloosas across the mountains on trails carved by wild animals,[89] where they met and became allies with the Flatheads and Crow on the Montana plains. (One path was the now-famed Lolo Trail, later used by explorers Lewis and Clark, and then by the Nez Perce in their attempt to flee the U.S. military.) There they started to hunt buffalo with these new allies. Hunting parties would spend months or even years following buffalo.[90] While they could not bring back enough dried buffalo to their territory to sustain their entire people, as their plains neighbors could do, they did bring back hides. And like their neighbors, they also adopted tipis and buffalo-hide clothing.

Nez Perce women had some traditional roles, such as gathering camas bulbs for food and caring for children, but they also skillfully tended and rode horses. Women were allowed to collect their own wealth and could also leave their husbands if they wanted.[91] These rights gave them a measure of power many white female settlers did not have at the time.

The Nez Perce gained a reputation for their careful breeding of horses with exceptional speed and stamina. Appaloosas became the envy of surrounding native peoples and white explorers, who later made their way through the

area. The distribution of these horses through trade, and sometimes raiding, served to improve the quality of the region's wild and domestic horses.[92]

Ironically, Nez Perce horses powered the early expeditions of white explorers and pioneers through the Northwest that paved the way for eventual American settlements in Nez Perce territory. These settlements later led to the loss of the Nez Perce's land and horse herds. On the other hand, the Nez Perce's expert horsemanship abilities and powerful Appaloosas gave them a fighting chance when they fled the U.S. military.

PART III

THE EAST
1500s-1700s

8

THE ROCKY ARRIVAL OF HORSES IN THE EAST

The horse's role in the earliest exploration of what would become the eastern U.S., with its swampy South and densely wooded North, was less significant than out West.[93] While horses were hugely beneficial to the exploration and settlement of those dry, exposed landscapes, they quickly became disadvantageous in the East.

Horses were impossible to ride through thick forests. And unlike in the West, they gave the Spanish colonists no advantage over the indigenous peoples of the Southeast, who were able to hide behind foliage and shoot horses as they were mired and useless in muddy terrain.[94] The conditions made early exploration of the coast by ship more effective than by horseback.

Three Spanish attempts in the 1520s to find gold and colonize the East—led by Juan Ponce de León, Lucas Vázquez de Ayllón, and Pánfilo de Narváez—ended in almost immediate failure due to the difficult landscape and Native American attacks.

Narváez, the *conquistador* who frightened the native peoples of Cuba into submission by galloping around on horseback ringing bells, found his techniques did not work out so well in Florida. Failing to find gold or food, he and his three hundred men were forced to eat their forty horses.[95] They proceeded to leave their animals' skeletons behind in what they called the Bay of Horses, near what is now Tallahassee. After their ship abandoned them, they attempted to build their own fleet and sail to safety in Mexico.

These ships sank in bad weather, and only a few men managed to swim ashore. Interestingly, one of those men, Mustafa Zemourri, an enslaved Moroccan, may have been the first Muslim to set foot in what would become the U.S.[96] While the mission was a miserable failure, Zemourri survived it and his further enslavement by local Native Americans that followed, eventually gaining his freedom and becoming a respected medicine man.

Just over a decade after the disastrous expedition of Narváez, as Coronado was traipsing across what is now New Mexico, Spanish explorer Hernando de Soto also tried his luck in the East. In 1539, he sailed from Cuba to Florida with around 230 horses to conquer the indigenous peoples and search for gold.[97] There he began his reign of terror across the Southeast, burning homes, ruining crops, and killing Native Americans. While the indigenous peoples he encountered were frightened of the horses, it did not help de Soto much. He could not gallop around intimidating them, because his horses sank deep into the swamps.

During this time, de Soto encountered the Chickasaw people, who would become significant horse breeders and traders in the years to come. Legend says the Chickasaw acquired their first Spanish horses from de Soto's expedition, but there is no evidence de Soto left any horses with them.[98]

Eventually de Soto made it as far west as the Mississippi River, where he died of illness. Only forty of the horses from his expedition were still alive by that time, but they too met their fates on the muddy banks of the river.[99] When Native Americans attacked, de Soto's remaining team fled to boats and watched them take down the mired horses with arrows.

The first English expeditions to the East did not fare much better, nor did they seem to benefit much from the use of horses. Between 1584 and 1590, the English made four voyages across the Atlantic to explore and colonize the Southeast coast.

The first voyage, led by Philip Amadas and Arthur Barlowe, was exploratory. They had been assigned by Sir Walter Raleigh to survey the coasts and scout potential locations for permanent settlement. In his trip reports, Barlowe reported friendly native peoples in what is now North Carolina. (Native Americans had inhabited North Carolina for about twelve thousand years prior to the arrival of Europeans.[100]) He also noted the wildlife encountered on North Carolina's Outer Banks. His list did not include horses.[101]

In 1585, the second voyage from England, led by Sir Richard Grenville, set out across the Atlantic with seven ships. Grenville stopped on Isabela, now the Dominican Republic, to trade for horses and other livestock.[102] He then sailed up the East Coast and left 107 men on Roanoke Island, a barrier island that is part of the Outer Banks, to start a colony under the leadership of Richard Lane.

The rest of the expedition's men divided up to continue exploring the coast, and Grenville sailed back to England a few weeks later to gather more provisions for the new colony. By the middle of 1586, the men on Roanoke were starving and likely ate all their horses and other livestock while awaiting rescue.[103] They jumped at the chance to sail home when an English ship eventually arrived and offered them passage, abandoning the colony.

It is unclear whether the third expedition from England in 1587 to set up a permanent colony, led by John White, brought along horses and other livestock when they settled once again on Roanoke Island.[104] White stayed long enough to meet his granddaughter, Virginia Dare, the first English baby born in the Americas. Soon after, he left the colonists to their own devices and sailed back to England to gather supplies for them.

But naval battling between the English and Spanish delayed White's return. In 1588, the English defeated the Spanish fleet of warships known as the Spanish Armada, which had limited England's seafaring capabilities. The victory restored England's freedom on the seas and opened up the East Coast to colonization by countries other than Spain, allowing England the opportunity to gain a foothold on the continent.

During the war, the Spanish captured White on the high seas. He eventually escaped and made his way back to Roanoke Island in 1590, on England's fourth voyage, to see what had happened to his family and the rest of the settlers. He did not report finding horses or other livestock on the island when he arrived.[105] Nor did he find the settlers. Referred to as the Lost Colony, their fate remains a mystery today.

For the next seventy-five years, the native peoples of the Outer Banks lived free of European settlement.[106] No evidence suggests horses survived on the islands during this time period.[107]

Horses offered little benefit to the early explorations and colonization of the Southeast by the Spanish and English. It was not until the Spanish established missions in the region that horses began to play a more significant role in life on the East Coast.

9

SPANISH MISSIONS AND THE CHICKASAW HORSE

As they did in the West, Native Americans played a significant role in the spread of Spanish horses throughout the East, particularly the Cherokee, Chickasaw, Choctaw, Creek, and Seminole. These nations were collectively and controversially referred to as the Five Civilized Tribes, because they eventually adopted cultural elements of the European colonists.

According to legend, the Chickasaw obtained their first horses from de Soto as he passed through Mississippi around 1540 on his violent escapades. But there is no evidence de Soto left any horses with the Chickasaw.[108] What we do know is that he and his men ran afoul of these native peoples during the encounter when de Soto demanded that two hundred of them become his porters. The Chickasaw beat them soundly, burned almost all their food, clothing, and supplies in the middle of the night, killed numerous men and horses, and sent the rest packing.

We also know the Chickasaw were an early nation to access the horse, whether from de Soto or the Spanish missions that cropped up in the South in the years following his expedition.

By 1615, the Spanish had established more than twenty missions across modern-day Florida and southern Georgia to lay claim to the land and subjugate Native Americans by converting them to Catholicism and putting them to work as indentured servants.[109] Missions were essentially fortresses meant to entice Spanish colonists to settle the land. Catholic priests ran them. Soldiers guarded them.[110] By 1650, Spanish settlement in the region had grown to seventy-two missions, eight large towns, and two royal haciendas.[111]

As in the West, these missions sustained themselves with farms and ranches, which included Spanish horses and other livestock. The Spanish had some luck in Florida and Georgia trying to subdue Native Americans who lived in permanent, agricultural communities under central leadership, using them as laborers. On these farms and ranches, the native populations learned horsemanship skills and gained access to Spanish horses.[112]

Missions did not use fencing, and horses lived freely on mission lands. Local Native Americans began to raid the horses, build up their own herds, and trade them to other indigenous peoples. By the early 1600s, these horses had spread north and west through the Cherokee, Chickasaw, Choctaw, Creek, and Seminole.[113]

Thirteen original British colonies, established by 1732, as well as eastern barrier islands with wild horse history.

JULIA SOPLOP

However, the horse did not entirely transform the cultures of these agriculture-based communities, as it did for the nomadic, buffalo-hunting Horse Nations.[114] These peoples primarily used horses as pack animals, as well as for some riding and racing.

The Chickasaw started to breed Spanish horses and became known for their high-quality, fast, hardy Chickasaw ponies. They traded their horses with English colonists for goods and weapons, spreading Chickasaw ponies up into the Carolinas and Virginia by the late 1600s.[115]

Soon, many English began to refer to any strong Spanish horse as a Chickasaw pony, regardless of its origin. Other names for this horse include Seminole pony, Florida Cracker, and Choctaw horse.

By the mid-1700s, the Chickasaw began to acquire horses from the Texas plains to increase their trading business. These horses, often driven from Texas, had either been caught from the wild, stolen from the Spanish, or bred by the Horse Nations.[116] The Osage, who lived at the edge of the plains, initially served as middlemen between the Horse Nations and the native peoples of the East, delivering the horses into the hands of the Chickasaw at the trading center of Natchez, Mississippi. The Chickasaw then spread these horses through Tennessee, Virginia, the Carolinas, Kentucky, and the Ohio Valley.

By the early 1700s, raids from English settlers of the Carolinas shut down the Spanish missions in Georgia and Florida.[117] But the horses the settlers and the Native

74

Americans obtained from those missions and spread throughout the colonies and into the wilds of Appalachia would fuel the colonies and become the foundation stock of the Quarter Horse, the most popular breed in the U.S. today.

10

EARLY HORSES OF THE COLONIES

E arly colonists of the East used horses to varying degrees, but most travel during this time period was done by boat, either by ocean, estuary, or river.[118]

In 1607, the English established their first permanent colony in Jamestown, Virginia. Twelve years later, those colonists purchased twenty or thirty enslaved Africans from English pirates, who had stolen them from a Portuguese ship after enslavers had kidnapped the Africans from what is now Angola.[119] Around four hundred thousand Africans would follow their path across the Atlantic Ocean to be sold into slavery in what would become the U.S. (About 12.5 million enslaved Africans were shipped to the Americas in total.) On their backs, the colonies would eventually grow into an independent nation and economic powerhouse.

During this period of the early 1600s, the infancy of the colonies and of African enslavement within them, English settlers did not see much value in horses.[120] Instead they preferred oxen, which could do farm work and hauling but also provide milk, meat, and hides. By 1625, the English had shipped over a few small groups of horses, but all of them died from Native American attacks or fell prey to starving settlers, who resorted to eating them in desperation.

After 1625, most horses shipped to English colonies were Irish Hobbies and Scottish Galloways, both strong and compact, averaging about 12-13 hands.[121] Hobbies were gaited, making for smooth riding horses. The Galloways had excellent cow-herding sense. These two breeds were tough and fast, like the Spanish horse.

Inland farms, too far from waterways to transport goods by boat, found horses advantageous during this period, but they were susceptible to Native American attacks.[122] Through the 1640s, brutal battles between indigenous peoples and farmers left many settlers dead. Their livestock died or scattered to the wild. Newly wild horses formed bands and began to spread through the Blue Ridge Mountains. By 1649, only two hundred documented domestic horses lived in Virginia.[123]

As more English began to pour into Virginia in the second half of the 1600s, these settlers spread toward the mountains looking for rich farm land.[124] They found the wild horses disruptive to their farming practices, so they captured and gentled some for their own uses. Others, they hunted down and shot.

But while they were trying to reduce the wild horse population in the mountains by removing and killing them, they were also contributing to the horses' spread. As in other regions, the early English colonists of the East did not often use fencing. So from Georgia to Maryland, the colonists' horses would escape and make for the Appalachian highlands.[125]

When the Spanish horses, referred to as Chickasaw ponies, made their way up to Virginia, colonists saw they were high-quality stock. They began to breed the Chickasaw ponies to their Irish Hobbies and Galloways, creating the little Virginia horse—the precursor to the Quarter Horse.[126]

Unlike the English colonists, the Dutch, who settled New York in 1624, typically used horses rather than oxen for farm work.[127] They imported large draft horses from the Netherlands, and used them to breed up, meaning to increase the size of, the Spanish horses they had accessed locally.[128] They kept herds on Long Island, out of reach of the native peoples.[129] The practice of keeping horses on islands or peninsulas, so as to require just one wall or fence to contain the horses, became common practice up and down the East Coast.[130]

In Pennsylvania, Dutch and German settlers also preferred draft horses for farm work and hauling goods to market, breeding up their own stock to meet these needs.[131]

At the same time, New Englanders did not see much use for the horse.[132] Puritans even found riding and horse racing sinful.[133] As late as 1632, there was only one documented horse in the Plymouth Colony of Massachusetts.[134] Within a few decades, however, New Englanders would understand the value of the horse and develop their own breeding industry.

Throughout the first several hundred years of horse ownership in North America, people often treated mares much like they often treated women: without much respect. The Spanish rode stallions, only using mares for breeding purposes.[135] This tradition continued in the Americas for centuries among many groups. Mares were thought of as weaker than stallions or geldings, and it was an embarrassment to ride them.

There was little understanding of the concept of heredity during the early colonial period. Breeders thought the stallion mattered most and often used local Spanish or English mares without much thought to their characteristics, other than shape and femininity, when pairing animals.[136] Mares were considered rather dispensable, and horse lineage was traced through the stallion's line.

All sorts of misconceptions about mares circulated among settlers, including that they would only produce offspring with the characteristics of the first stallions with which they bred.[137] It would not be until the early to mid-1900s, when the scientific community developed an increased knowledge of genetics, that breeders began to understand the significance of the mare.

11

WILD HORSES OF THE BARRIER ISLANDS: LEGEND AND HISTORY

When it comes to the origins of the wild horses of the eastern barrier islands, legends have become so interwoven into regional identities one could argue that whether this lore is accurate now takes a back seat to the fact that its residents have been retelling the stories for centuries. The legends themselves have become a part of history and culture, and yet the exact origins of the horses still remain a mystery.

Interestingly, many of these barrier island legends are similar, especially in Maryland, Virginia, and North Carolina.

In her 1947 novel, *Misty of Chincoteague*, Marguerite Henry catapulted the wild horses of Maryland and Virginia's Assateague Island into the national spotlight, along with the legend that they swam ashore from a shipwrecked Spanish galleon. One historian suggested that while a shipwreck could have been the origin of the island's first horses in the late 1600s or early 1700s, a Spanish ship carrying horses from Spain would not have been in these waters at that time.[138] It would have been more likely that an English or Dutch ship had picked up Spanish horses in the West Indies and wrecked while making its way up the coast.[139] Another author asserted the horses actually came from a Spanish galleon, *La Galga*, that was blown off course by a hurricane and sunk in 1750 off the coast of Virginia.[140]

The most plausible scenario for the first horses on Assateague, however, is that in the late 1600s, English colonists who made their homes on Chincoteague Island used neighboring Assateague to pasture their horses, eventually abandoning some of them.[141] (Assateague and Chincoteague are next to each other, but the wild horses live on Assateague.) No evidence suggests the settlers found horses already living on Assateague when they arrived.[142]

Local lore suggests that beginning in the 1700s, residents would swim the horses across the channel annually from Assateague to Chincoteague to capture, brand, and sometimes remove unclaimed horses to sell or use on their farms.[143] The first documentation of this practice on the islands, known as pony penning, was in 1835. In 1925, the Chincoteague Volunteer Fire Company, which now owns the herd, took over the annual pony penning as a fundraiser. "Salt water cowboys" would round up the horses on Assateague and swim them across the channel to Chincoteague at low tide, auction off the foals, and swim the rest of the herd back to Assateague. The event continues to this day, still serving as a fundraiser to support the horses' management.

The wild horses of Assateague are significant culturally and historically, both as legends and as the animals that powered many early farms on the Virginia coast. In more recent years, they have also served as important research subjects for population management techniques, as we'll discuss later.

South of Assateague, North Carolina's Outer Banks have been home to wild horses, known as banker ponies, for several hundred years. The origins of these Spanish horses are similarly murky, and the legends familiar: the horses swam ashore from sinking Spanish ships or were left behind from the early colonization attempts of the 1500s.[144] As we've discussed, however, there is no evidence of horses surviving England's earliest colonization attempts.

The shipwreck explanation is feasible; the waters off the Outer Banks are incredibly dangerous due to turbulent seas and shallow shoals. This area, known as the Graveyard of the Atlantic, is the location of at least a thousand shipwrecks.[145] During the colonial period, horses did wash off decks in storms here.[146] Ships did sink with horses on board. Shipwreck debris did wash up on the Outer Banks. But evidence does not exist that the horses actually survived, reached the islands, and thrived. Still, locals have passed these legends down for hundreds of years, making the romantic stories themselves part of the islands' cultural heritage.

What we do know is that by the 1650s, settlers who had moved down into the lowlands north of the Albemarle Sound to escape persecution and upheaval in the Virginia colony were using the Outer Banks to pasture their horses.[147] This region hosted a diverse mix of people, most of whom were small-time farmers: Quakers seeking religious freedom, formerly enslaved peoples, runaways from the law, as well as Native Americans.

A wild stallion braces as a storm barrels over the dunes on North Carolina's Outer Banks.

In 1663, North Carolina became a separate colony from Virginia, and colonists began to settle on the barrier islands.[148] More farmers also started to leave their horses and other livestock on the islands to avoid the British tax on fencing, first imposed in 1670. Within a decade, locals were placing stock animals, including horses, on the northern Outer Banks—the location of today's Corolla wild horses, managed by the Corolla Wild Horse Fund. The Spanish horses of these colonists flourished on the islands, foraging on marsh grasses and digging for fresh water. A recent genetic analysis of three Outer Banks horse herds—Corolla, Shackleford Banks, and Ocracoke—have confirmed their shared origins as colonial Spanish horses.[149]

Annual or bi-annual pony penning roundups on the

Wild horses on the Outer Banks seek the offshore breeze to keep the flies away.

Outer Banks were well-established by the 1700s.[150] Stockmen would collect, divvy up, and brand horses, taking some to the mainland to sell or use on their farms, and leaving others to reproduce. Eventually, they abandoned the island herds altogether.

The Sea Islands of South Carolina and Georgia were also home to wild horses, which can still be found today on Georgia's Cumberland Island. Legend holds that the Spanish set their horses free after colonizing this island from 1566 to 1675, though there is no evidence to support this theory.[151] Another possibility is that the English brought horses to the island when they built a fort on it in 1736, but there is no documentation of wild horses there until 1788.

Regardless of their origins, the Spanish horses of the Sea

Wild horse of Spanish origin on an eastern barrier island.

Islands, called Marsh Tackies, have played numerous roles in regional history, as you will read later. Marsh Tackies, known for their ability to navigate swampy lands that sink and spook most horses, crop up in significant stories of the Revolutionary War, the Civil War, the Reconstruction Era, Gullah culture, and World War II.

The origins of the Spanish horses of the barrier islands of Maryland, Virginia, North Carolina, South Carolina, and Georgia, may be unclear, but two things are certain: these horses played a variety of roles in the history of the region, both in its settlement and in powering its farms; and the legends themselves surrounding their arrivals on the islands have secured a firm place in the area's cultural record, regardless of their veracity.

These herds also throw a twist in the debate of whether modern mustangs in the U.S. are native versus nonnative, wild versus feral. Unlike the western U.S., which is packed with fossil evidence of the horse's evolution there, the eastern barrier islands have no fossil evidence of ancient horses. While the horse originated in the U.S., it did not co-evolve with a barrier island habitat over the course of millions of years. And yet, we know horses have now been part of the islands' ecosystems for several centuries. Today, the various agencies that manage the remaining wild horses on these islands classify them in different ways, which informs their management policies.

12

EARLY HORSE RACING AND BREEDING IN VIRGINIA

Humans have likely raced horses almost as long as they have been domesticating them, around five thousand years. Horse racing in England dates back to around 200 CE. When English subjects sailed to the colonies, they carried their passion for racing with them.[152] But the dense woods of the East Coast made it nearly impossible for early colonists to carve out and recreate the lengthy, oval racetracks popular in England at the time.

In Virginia, all it took to initiate a horse race was a couple of work horses. A straightaway of about a quarter mile would do, whether down a road or through a trail in the woods.[153] Race spectators from every echelon of society gathered and placed bets, but only "gentlemen," as the colony considered wealthy white men, were allowed to enter their own horses in a race. However, race jockeys in Virginia during this period were often enslaved Africans, as were horse trainers and stable hands—skilled trades that would influence the status of these men following the abolition of slavery.

Virginia men first raced their work or riding horses, mostly Hobbies and Galloways, which were fast at racing the quarter mile. Then Chickasaw ponies made their way to Virginia and turned out to be even faster at this distance. The colonists bred the Chickasaw ponies with their Hobbies and Galloways and realized the combination of these types made the fastest horse at the quarter mile. Here, the original Quarter Horse was born. This compact horse, typically fewer than 13.2 hands, was not just speedy on the race course. It was also practical for daily life and work in the region.

Using enslaved laborers, the owners of tobacco plantations in Virginia grew wealthy during the 1600s and 1700s. They had a passion for horses, for wagering, for winning big. And now they had money to burn.

Virginians became more interested in long-track racing and eventually started to build oval race tracks as they cleared the land. As early as 1730, they began to import horses specifically for racing, purchasing Thoroughbreds — hot-blooded, long-race horses with Arabian, Spanish, and Galloway blood — from England to breed their own American Thoroughbreds and early Quarter Horses.[154]

In 1756, an English Thoroughbred stallion named Janus arrived in Virginia. When Janus bred with Virginia mares, which were combinations of Galloways, Hobbies, and Chickasaw ponies, the offspring were magnificent.

Supposedly, Janus contributed initial speed and refined looks to these racers.[155] His offspring essentially solidified the early Quarter Horse, which gained a strong reputation not only for racing but also for pulling and riding.[156] Later, a Thoroughbred named Sir Archy contributed endurance to the line.[157]

By this time, Virginia had developed a horse culture. The horse became not just a matter of practicality in the colony but a status symbol. Unlike the Horse Nations, though, Virginians' status did not particularly depend on the number of horses a person owned, but rather whether a horse was high-bred, fashionable, expensive to acquire, a winner of races, and perhaps a skilled fox hunter. Virginia's colonial-era horses would influence numerous American breeds and provide much of the cavalry for both sides of the Revolutionary and Civil Wars.

13

UP NORTH: THE NARRAGANSETT PACER AND CONESTOGA HORSE

Demand for horses throughout the colonies differed widely depending on the terrain and settlers' livelihoods. New Englanders, who did not initially find much value in the horse, began to see its usefulness several years after settlement in pulling stumps and clearing the densely wooded lands of the Northeast for farming.[158]

Early settlers began to use their horses as pack animals to transport goods, as well, but rarely rode the first horses they imported due to the difficulty of the terrain and lack of roads in the region.[159] Unlike wealthy Virginians, most New Englanders at the time were not affluent, so instead of purchasing specialized horses for different tasks, settlers required all-around work horses.

Based on the desire for a flexible animal and a more comfortable saddle horse to handle rough terrain, the Narragansett Pacer, often considered the first truly

American breed, emerged in Rhode Island in the late 1600s.[160] The origin of the breed is not certain, but it likely arose from a combination of English, Dutch, and French-Canadian horses.[161] It was small at about 14.1 hands, typically chestnut/sorrel in color, and its smooth gait made for a comfortable ride across the difficult landscape of the Northeast.[162] It also served as a great horse for harness racing and soon became extremely popular in the colonies.

By the end of the 1600s, Rhode Island had emerged as a breeding center for the Narragansett Pacer and started to export the horse by ship to other colonies, as well as to the West Indies to work on sugar plantations. The Pacer spread throughout the colonies and later influenced many gaited American breeds. George Washington owned two Narragansett Pacers, and legend has it Paul Revere made his famous ride of the Revolutionary War on a Pacer.[163]

But by around 1800, the Narragansett Pacer itself had disappeared. The working conditions on the West Indies sugar plantations were terrible and usually resulted in early deaths of the horses.[164] It is likely the demand for Pacers on the sugar plantations was so high they were eventually all exported. In their place, larger trotting horses became the rage.[165]

In the early 1700s, the Pennsylvania Dutch needed to transport large quantities of goods from Lancaster, Pennsylvania, to Philadelphia to feed the city.[166] To do so, they developed the heavy Conestoga wagon, which was massive and had large wheels and a canvas cover

HORSE GAITS	
Natural gaits	
Walk	Four-beat sequence: left hind; left front; right hind; right front.
Trot	Two-beat sequence of diagonal pairs: left front and right hind touch ground simultaneously, followed by a moment of suspension, then right front and left hind touch.
Canter	Quick, three-beat sequence that feels like a rocking horse to the rider: outside hind touches ground, inside hind and outside front join outside hind on ground; outside hind picks up while inside front hits ground; inside hind and outside front pick up; inside front picks up (all four legs off ground). A slower version of the canter used in western riding is called the lope.
Gallop	A four-beat sequence, during which all four feet hit the ground in rapid succession in the same order as the canter, followed by a long moment of suspension.
Specialized gaits (natural or artificial)	
Pace	Two-beat sequence that makes for a comfortable ride at a slower pace: legs move in lateral pairs; two feet are always off the ground. Harness racers are often natural pacers.
Amble	Encompasses several smooth, four-beat gaits that are faster than the walk but often slower than the canter. Includes the running walk natural to the Tennessee Walking Horse, and the American Saddlebred's trained rack.

reminiscent of wagons in Europe during the Middle Ages.

To haul the wagon, they needed heavy horses. So they developed the first American draft breed, the Conestoga horse. We know little about the Conestoga horse except that it was a medium-sized draft, roughly 16-17 hands and around sixteen hundred pounds.[167] A Conestoga wagon, which may have been the most efficient of its time, required six to eight Conestoga horses to pull it. [168]

By the middle of the 1700s, there were an estimated seven thousand Conestoga wagons throughout Pennsylvania.[169] They became even more popular once the country's first engineered road, the Philadelphia and Lancaster Turnpike, opened in 1795. (This toll road was called a turnpike because the operator would swing open a "pike" across the road after receiving payment.) Drivers drove this wagon on the right side of the road, supposedly starting the tradition of right-side driving in the U.S.

Conestoga horses and wagons grew immensely popular. The wagons were eventually transformed into prairie schooners, the type settlers and traders drove along the Oregon and Santa Fe Trails.[170] These pioneers often used oxen to pull their wagons instead of horses, however.

The Conestoga horse disappeared altogether by the mid-1800s, pushed out of popularity by the arrival of larger draft breeds.

14

SIGNIFICANT BUT NOT-SO-FAMOUS RIDES OF THE REVOLUTIONARY WAR

A messenger arrived after dark one night in 1777, desperate to alert the local militia that the British were attacking a nearby town. But, as was the custom during the Revolutionary War, the militia of four hundred men had temporarily disbanded to go home and plant their crops. The messenger was too exhausted and unfamiliar with the area to continue, so another rider took his place.

Setting out around 9 p.m., this expert rider rode nearly forty miles through the rainy night, navigating heavily wooded, roadless terrain and carefully avoiding bandits and British loyalists, to alert the militia that they needed to gather and fight off the British.[171] The rider successfully summoned the men, who traveled to the town under siege. Although they were late to the battle and too outnumbered to win, they managed to put up a good fight against the British soldiers as those soldiers retreated.

The rider's contribution to the war effort was considered so important the rider earned a personal thank you from General George Washington.[172]

Sound familiar? You've probably heard of Paul Revere's famous midnight ride. But this rider was not Paul Revere. She was a sixteen-year-old girl named Sybil Ludington, who rode twice as far as Revere did, through much more complex and dangerous conditions.[173] While some historians question the veracity of details of both these rides, the rides are largely considered more historical event than legend. They both demonstrate, however, that the lines between history and legend are often blurred, the details lost to time.[174]

Sybil Ludington was not the only expert female rider to make a significant contribution to the Revolutionary War effort, at least according to legend. The story of Betsy Dowdy has been told on North Carolina's Outer Banks for centuries. Like the wild horse lore, this legend has earned its own place in cultural history.

According to the story, when the Revolution began, sixteen-year-old Dowdy lived on the Currituck Banks, the northern portion of the Outer Banks where the Corolla wild horses still live today.[175] In 1775, an acquaintance arrived by boat, warning Dowdy's father that the British Lord Dunmore was assembling his men nearby and planned to raid the barrier islands of horses and supplies. Only General Skinner, the leader of a local militia, would have the power to stop the British advance, and his camp was fifty miles away.

Dowdy's father felt it was too late to try to warn Skinner. Betsy Dowdy disagreed. She snuck off on her banker pony, Black Bess, with nothing but a knife for protection against the night. She and Black Bess swam across the Currituck Sound to the mainland, where Dowdy rode the fifty miles through marshland and forest, convincing a ferry operator to give her a lift across the Pasquotank River to reach Skinner's camp. Her warning resulted in the militia halting the advance of the British—an important victory that showed the Continental Army (the Americans) could, in fact, defeat the British.

While evidence supports Betsy Dowdy's existence and the historical event in which she purportedly played a role, evidence does not exist to prove or disprove her ride.

Since the dawn of documentation, historians, almost exclusively male for much of civilization, have systematically left out of recorded history contributions of women and people of color, downplaying or ignoring them outright in favor of remembering and celebrating the accomplishments of white men. Although different schools of thought have dominated historical analyses throughout each era of U.S. history, it was not until the 1960s and 1970s, amid the Civil Rights and feminist movements, that more historians began to focus on social history, meaning broadening the scope of study to include the lives of people such as women and people of color rather than simply the elite, white men who have dominated the historical narrative.[176]

Many women throughout history were every bit the expert riders their male compatriots were. It just takes more research to uncover their stories, along with an acceptance that the details of those stories may never fully emerge due to limited documentation.

15

THE REVOLUTIONARY WAR
AND AFTERMATH

History books sometimes gloss over the fact that it took
another 125 years after the founding of Jamestown in
1607, along with many skirmishes and treaties with other
nations, for the British to establish their original thirteen
colonies along a thin strip of the East Coast.

As the 1700s unfolded, the colonies developed their own
unique populations and ways, remaining rather isolated
from one another. But it became apparent over time that it
would benefit them all more to band together and strike out
on their own than to remain under British rule.

Beginning in 1775, the colonies spent eight years fighting
off Britain during the Revolutionary War. (It was not
officially considered a war against the British until the
following year when the Second Continental Congress
issued the Declaration of Independence to establish the U.S.
as its own nation.) The war both used and depleted much of
the colonies' horse population, and also set the stage for the

expansion of the horse's role in the U.S. following the war.

The British had a significant advantage in military strength, experience, and training over the Continental Army, which was comprised mostly of a hodge-podge of farmers ill-trained for battle. But Britain's expertise lay in the traditional siege techniques of old Europe; they were less prepared for the varying forms of war, often involving the skillful use of the horse, that the Continental Army would wage against them.

General George Washington, a renowned horseman, and other Continental Army strategists did not think cavalry would be particularly helpful in the densely wooded Northeast.[177] So in the early years of the war, they primarily used horses as mounts for commanders and scouts, and as pack animals to pull artillery and supplies.[178]

Soldiers also used horses to come and go from the war. As we discussed in the account of Sybil Ludington's ride, soldiers would ride home to tend to their farms during planting and harvesting seasons, then return to their militias and take up the war effort again. [179]

Meanwhile, when the British landed in large port cities, they would send out men to either buy horses, if they were in a place loyal to the British Crown, or raid the enemies' horses.[180] In New England and Pennsylvania, they acquired draft horses common to the regions, while in the South, they picked up lighter horses to use as officer mounts and cavalry.

Both armies raided Virginia horse country for high-quality horses, especially for Chickasaw ponies, Quarter Horses, and Thoroughbreds.[181]

Locating and transporting forage for horses was a factor that limited their use in both militaries.[182] In fact, the armies spent much of the war destroying farms and store houses to prevent the enemy from feeding their horses.

However, as the war dragged on, cavalry began to play a larger role for the Continental Army in the South. While the British moved through the region on foraging missions, stripping it of horses and food, the Continental Army began to retaliate by sending out mounted soldiers, many of them skilled Virginia horsemen, to launch surprise attacks.[183] They did not battle on horseback but used their horses to move quickly across territory, position themselves to strike, and make rapid escapes back into the wilderness.[184]

In the swamps of South Carolina, the Marsh Tacky, the little Spanish horse of South Carolina, Georgia, and Florida that handled well and did not spook in the Lowcountry marshes, played an outsized roll in the guerilla fighting.

Controversial South Carolina militia commander Francis Marion earned his place in history and his nickname, Swamp Fox, thanks to his Marsh-Tacky-riding militia's ability to attack and outmaneuver the British cavalry, who rode heavier horses unaccustomed to the boggy conditions.[185] Marion had learned by observing the Cherokee, against whom he had fought during the French and Indian Wars, how to blend into his surroundings and use the terrain to his benefit.[186]

With this rag-tag militia combined with his swamp prowess, Marion successfully launched guerilla attacks throughout South Carolina, including one incident when he and fifty of his men ambushed a British camp and rescued 150 American soldiers.[187] His tactics helped turn the tide in favor of the Continental Army at a crucial point in the war.

The birth of the U.S. made way for numerous changes in the use of horses by the end of the 1700s. First came the consequences of the Revolutionary War itself. Although far fewer horses were used during the Revolution than would be used several decades later in the Civil War, the numbers of high-quality horses acquisitioned, stolen, and killed by both militaries diminished the quality of stock in some areas, forcing breeders to try to rebuild their lines.

Virginia was particularly hard hit during the war, because it was a breeding center known for superior horses. The loss of horses to the war, along with the lack of importation, racing, and breeding during the war, left the state reeling.[188] The great reduction in the quality of remaining horses left in Virginia, coupled with the number of people flowing westward into Kentucky and Tennessee with their Virginia horses to find new lives on the frontier, meant the state began to lose its dominance in breeding, an important part of its economy.

Another consequence of the war was a welcome one: the cutting of more roads. Prior to the war, waterways provided most travel routes within or between colonies.[189] But as the militaries moved their troops and supplies throughout the East Coast during the war, they carved rudimentary roads through the thick forests, opening up new routes.

These vestiges of war, combined with a new desire for faster, reliable transportation between the newly unified states, led to the construction of more and better roads to connect major cities. And with new roads came a need for more carts and carriages for a variety of uses, as well as more horses to pull them. By the end of the 1700s, the East was primed for an explosion in the use of the horse for longer-distance travel and transport.[190]

The first president of the U.S., George Washington, worked to improve the country's agricultural production.[191] He had heard that mules, the typically sterile offspring of horses and donkeys, were excellent work animals. They were strong, could subsist on poor-quality feed, and could endure abuses common on plantations that horses could not, sometimes living into their forties.[192]

Shortly after the Revolutionary War, Washington acquired a few donkeys from Spain and Malta to begin breeding mules.[193] There had been some mules in America since the Spanish arrived, but they were not yet common. Washington quickly saw the value of the mule was real and began to promote it as a work animal throughout the states.

The popularity of the mule exploded after Eli Whitney's 1793 invention of the cotton gin, a device that separates cotton fibers from their seeds significantly more efficiently than can be done by hand, making cotton extremely profitable.[194] The South took note, quickly replacing other crops with cotton and developing itself into a premier cotton-growing region. Soon cotton plantation owners preferred mules to horses or oxen for the often brutally heavy work they demanded, although they still rode horses to manage their operations.[195] Meanwhile, the North continued to use horses for most farm labor.

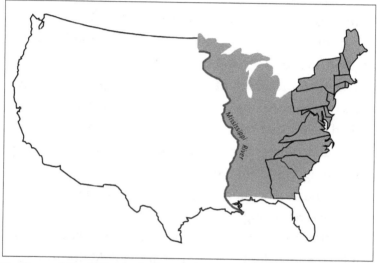

Territory of the U.S. (shaded) as defined by the 1783 Treaty of Paris, which ended the Revolutionary War.

As a result of the Revolutionary War, the U.S. not only gained its independence from Britain but also expanded its land holdings from the thin strip of colonies along the coast all the way to the Mississippi River, excluding Florida and the northern coast of the Gulf of Mexico. Settlers began to push west, chasing the country's new frontier. The government expanded roads in an effort to unify once-disparate settlements. New roads led to the expansion of travel and trade throughout the country, all powered by the horse. Whereas the horse had once been questionable in its necessity in some regions, it quickly became essential to productive living in the early years of the new nation of the U.S.

PART IV

THE WEST
EARLY TO MID-1800s

16

EXPLORERS, TRAPPERS, AND PIONEERS

J ust twenty years after the Treaty of Paris ended the Revolutionary War and drew the western borders of the newly formed U.S. at the Mississippi River, the country acquired the Louisiana Territory from France. This territory had bounced between Spanish and French control, in name anyway, throughout the 1700s. The 1803 agreement, known as the Louisiana Purchase, doubled the size of the U.S. The country now extended west from the Mississippi River across much of the Great Plains, excluding Texas, to the foot of the Rocky Mountains, lands inhabited by the Horse Nations.

President Thomas Jefferson hired Merriweather Lewis and William Clark to lead an expedition to scout the new territory, locate the Northwest Passage—a mythical and, as it turns out, non-existent waterway leading all the way to the Pacific Ocean—and develop positive relationships with the Native Americans they encountered. They were also tasked with recording everything they encountered: landscapes, weather, plants, animals, people.

Two of the most valuable members of Lewis and Clark's team were not the mountain men they had painstakingly assembled, but rather a black man enslaved by Clark named York, and the teenage Shoshone bride of a French trapper they had employed. The woman, Sacagawea, gave birth to a baby boy during the expedition and carried him on her back the rest of the way while playing the critical roles of wilderness guide and negotiator with the indigenous peoples they encountered.

In 1804, the expedition set out by boat up the Missouri River. When they reached Montana and encountered the Bitterroot Mountains of the Rockies, their hopes of a Northwest Passage were dashed. They realized that without horses to cross the mountains with their heavy supplies, they would be forced to turn back and abandon the rest of the journey.

Soon they encountered a band of Shoshone led by Sacagawea's brother. Despite the fact that the Blackfoot had just stolen a huge number of the Shoshone's horses, making her brother hesitant to give up any animals, Sacagawea was able to negotiate for twenty-nine scrubby pack horses in exchange for some supplies and the promise of white men bringing the Shoshone guns after the completion of the expedition.[196] This negotiation allowed the expedition to successfully cross the mountains by way of the Lolo Trail.

The expedition later had a positive encounter with the

Nez Perce, and Lewis recorded in his journal his esteem for the high quality of their Appaloosas and even their gelding methods: "...the other horses which we casterated are all nearly recovered, and I have no hesitation in declaring my beleif that the indian method of gelding is preferable to that practiced by ourselves."[197]

The Nez Perce agreed to take care of the expedition's horses for the winter while the team continued their journey by boat through the Pacific Northwest.[198] On the return trip east, the expedition picked up their horses and used them to travel as far as the Yellowstone River, where a group of Native Americans promptly stole the animals.

Although scattered French-Canadian fur trappers, known as mountain men, had traveled this area before Lewis and Clark using Native American hunting trails, the maps and records the expedition produced ignited further white exploration of the region.

More trappers and explorers began to head west, searching for beaver pelts and easier passages through the mountains.[199] Early trappers often traveled alone with two horses: one to ride and one to carry supplies and pelts.[200] They would stay with Native Americans they encountered and trade with them for fresh horses. As the fur business grew, companies bought large numbers of horses from the indigenous peoples of the Pacific Northwest to outfit their trappers. The trappers often lost horses to Native American horse raiding, as well.[201]

Soon trails first used by Native Americans and then

trappers and explorers became gateways to the West for white settlers. In 1836, the first wagon train of missionaries started west from Independence, Missouri, to Idaho. Within a few years, the trail would extend as far as the Willamette Valley of Oregon, which offered settlers excellent land for farming. Known as the Oregon Trail, it would serve four hundred thousand pioneers on their westward journeys before the transcontinental railroad made it obsolete.

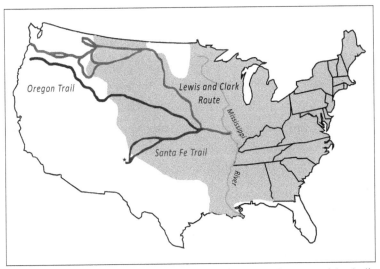

U.S. territory after the Louisiana Purchase of 1803 (shaded). Approximate trails mark Lewis and Clark's route, the Oregon Trail, and Santa Fe Trail.

Although the pioneers typically used oxen to pull their prairie schooners—that is, Conestoga wagons outfitted for family travel—horses were an important part of the journey. [202] They were used for driving cattle, hunting, and guiding.[203] Settlers would trade for fresh Native American horses along the way and arrived with these horses in the Willamette Valley to begin their new lives. They often referred to the horses they acquired from native peoples as "cayuse," regardless of whether they came from the Cayuse nation. The name became a derogatory term for these often-small-statured horses, the settlers not understanding the animals' sturdiness or stamina. The name also outlived the Cayuse people themselves, who died out by the mid-1800s from diseases whites settlers had spread along the trail.[204]

While many of the encounters between pioneers and indigenous peoples along the Oregon Trail were peaceful, the travelers did lose animals to Native American horse raids.

In 1821, men on horseback driving a pack train of mules blazed a more southerly route west, known as the Santa Fe Trail.[205] It connected Independence, Missouri, with Santa Fe, New Mexico. Mexico had just taken New Mexico from the Spanish, and Americans used the Santa Fe trail as a commercial route to transport supplies to sell to the Mexicans.[206] Again, oxen were typically used to haul wagons, but horses were an integral part of the journey, just like on the Oregon Trail. Each wagon train brought along hundreds of horses to trade, which they exchanged, along with other supplies, for silver coins and mules to use on southern sugar and cotton plantations.[207]

The Santa Fe Trail passed through Comancheria. While the Comanche were still actively raiding horses throughout

Texas and Mexico, they mostly stayed clear of the wagon trains in the early years. There were occasional reports of horse raiding along the trail from the Comanche and other Horse Nations, but the Comanche were less concerned with people passing through to trade than settlers taking their lands.[208]

Horses played essential roles in the surveying and settlement of the American West. From the famed early explorations of the expanded U.S., to the fur trappers roving the Rocky Mountains, to the wagon trains of pioneers and traders searching for new land and commercial opportunities, horses allowed these groups to survive, travel through, and conduct business across the country's newly acquired, unfamiliar, unforgiving terrain.

17

RANCHES OF SPANISH CALIFORNIA

Due to its isolated geography and unique landscape, California has its own equine history, which does not dovetail much with the rest of the West until the Gold Rush years of the mid-1800s.

The Spanish, by way of Mexico, began to colonize California in earnest by 1769, building a chain of coastal missions from San Diego to San Francisco.[209] As it did with its missions across the Southwest and Southeast, Spain constructed these California missions to gain a foothold on the land before other European countries did, as well as to convert and subjugate the Native American population.[210]

The Spanish and other Europeans brought with them diseases that nearly wiped out California's native peoples. Before the arrival of the Spanish, an estimated three hundred thousand Native Americans lived in California.[211] Within a century, that number plummeted to around thirty thousand.[212]

The missions depended on their own farms and ranches to provide food and income through trade. They also depended on the forced labor of the Native Americans to build and work these ranches, where they raised livestock, including cattle, sheep, and horses.[213]

Horses were essential to mission life in California. The land between the coast and the mountains was lush with grass and provided excellent grazing for livestock.[214] Like other communities at the time, the California missions did not use fencing. Horses left on their own quickly multiplied in this environment, and many became wild. By 1800, there were an estimated twenty-four thousand horses in California.[215]

Sometimes the ranches shipped horses and cattle to Hawaii and other locations to sell.[216] But their principle commodities were cow hides and tallow (fat), which meant one of the jobs of the cowboys was to slaughter thousands of cattle each year, often accomplished on horseback. One technique was to ride a horse up alongside a cow and either break its neck with one swift blow or cut its hamstring with a long knife.

Ranches sometimes used wild horses to thresh grain— that is, to separate and prepare it for use.[217] They would pile an entire harvest in a corral, then herd in up to sixty wild mares to trample it for a few hours. The horses were often injured in the process.

The Spanish in Mexico had been ranching for a century by this point, and the settlers in California had adopted many of the techniques and gear of Mexican cowboys, called *vaqueros*, to manage their cattle.[218] The Spanish lived in the saddle as they moved their cattle around their immense pastures, collected calves for branding, and separated cattle for sale.

The cowboys used horses for entertainment, as well. Some of the most daring of them with the steadiest horses would rope grizzlies on horseback—two men at a time, since a roped bear could easily charge a single horse and rider.[219] They would choke the bear to the ground, drag it to a corral, tie it to a wild bull with a thirty-foot rope, and watch with a crowd of onlookers as the animals fought to the death. The bear usually won.

It is difficult to imagine today, but in the early 1800s, the Los Angeles area was horse country.[220] A horse culture emerged there among the Spanish settlers that was more reminiscent of the Horse Nations than any other Spanish colony that would later become part of the U.S.[221] Children grew up on horseback, and work, entertainment, and art— such as weaving horsehair—revolved around the horse. [222]

Unlike in other areas of the West, California's isolated, high-quality pasture lands actually produced more horses than the settlers had use for. These California ranches were separated from the Southwest by rugged mountains, scorched desert, and sometimes hostile bands of Native Americans. By 1805, since geography limited their trading, the Spanish had resorted to slaughtering wild and domestic horses by the thousands when their numbers got so high they were eating grass needed for cattle.[223] Ranchers would run them off cliffs or round them up, send them through a narrow gate, and slash them with a lance that ended in a crescent-shaped blade.[224] Some reports suggest they killed up to forty thousand per year, often without even retrieving their hides to sell.[225]

In 1833, more than a decade after Mexico won its independence from Spain, it began to shut down the mission system and distribute mission lands in California to Mexican settlers.[226]

While these settlers did not show much interest in herding their cattle eastward to sell, some Americans trekked west on what became known as the Old Spanish Trail to trade for horses to bring back to Santa Fe to sell.[227] They and local indigenous peoples stole many horses along the route, too.[228] Horses that escaped from trading and raiding along the trail are thought to be the origins of Utah's Mountain Home Range wild horses.[229]

California had more horses than it could possibly use for about the first half of the 1800s, and, like the Horse Nations, the Spanish and Mexican settlers there developed a culture that revolved around the animal. The California Gold Rush would change everything.

18

MINING

G old transformed the horse country of California perhaps more rapidly and irreversibly than any location in the West. In 1848, gold was discovered near Sacramento just days before the U.S. signed a treaty with Mexico to end the Mexican-American War and officially claim the territory.[230]

At that time, California's nonnative population was fewer than a thousand people and its native population was around 150,000—half what it had been before the Spanish arrived.[231]

But the California Gold Rush had officially begun, and people started to pour into the region. Americans came by overland trails, including off-shoots of the Oregon Trail, hoping to strike it rich. People from countries all over the world arrived by ship with the same goal. By the end of 1849, the nonnative population of California had exploded to one hundred thousand.[232] These migrants became known as the '49ers.

Prior to the Gold Rush, California ranches had more horses than they could ever use. But suddenly there was skyrocketing demand for those horses. The mines and mining camps that sprang up, as well as lone prospectors who wandered the mountains, depended heavily on horses, burros, and mules. Profitable minerals often lay buried in rough mountain terrain, well off railroad lines and maneuverable rivers, so mining camps relied on horses for supplies, transportation, and mail.[233] Horses also hauled the mined minerals to market.

For the first time, some traders were even turning around and bringing horses from Santa Fe to California to sell, instead of the reverse.[234]

The California Gold Rush peaked in 1852, before surface gold became scarce, but continued through the end of the decade. By then, the nonnative population of California was 380,000.[235] A decade or two later, the Native American population had fallen to just thirty thousand, mostly due to diseases nonnative settlers had brought.[236]

CALIFORNIA POPULATION SHIFTS DUE TO GOLD RUSH		
Year	Native Pop.	Nonnative Pop.
1768	300,000	0
1848	150,000	< 1,000
1870	30,000	380,000

Long gone was the California sprinkled only with Native American villages. Gone too was the dense horse country that awaited miners just fourteen years earlier. By now California had established towns and more infrastructure, as well as demand to meet the surplus of horses that had existed since the arrival of the Spanish. Although gold mining began to wane, settlement continued.

Gold rushes and mining operations outside of California, all sustained by the horse, created white settlements throughout the mountain West. These settlements had an enormous impact on Native American populations, including the Horse Nations of the Pacific Northwest. Purchasing horses from nearby Horse Nations became an essential way for mining operations to supply their camps.[237] And when whites found gold on native lands, they systematically pushed nations off those lands to access it. In a later chapter, we'll discuss how gold discovered on Nez Perce lands affected their future.

19

EARLY TEXAS COWBOYS AND COW PONIES

F ew historical figures have captured America's adoration like the image of the early cowboy and his horse—a romanticized and enduring symbol of the West that has invited endless representation in literature, lore, and film.

The peak of long cattle drives across a truly open range came after the Civil War and only lasted a few decades before the Great Plains were divided up and fenced off. But the American cowboy's story in the West begins earlier.

By the late 1700s, after failing to convert the Apache, Spanish missionaries had hightailed it out of the Nueces Valley of Texas, abandoning their longhorn cattle as they left.[238] The cattle became wild and multiplied, but rivers mostly contained their herds within the valley.

After Mexico threw off Spanish rule in 1821, it began to invite American settlers into Texas, in large part to act as a buffer between Mexican settlements and the raiding

Comanche.[239] (The U.S. did not acquire Texas until 1845.)

While traditional histories often recount white American settlers trickling into Texas during this time, they neglect an important fact: many of these white settlers brought enslaved blacks with them. By 1825, these enslaved peoples made up about a quarter of the American population in Texas and had become an integral part of the western cattle industry.[240]

American settlers, almost as soon as they arrived in Texas, began to experiment with rounding up the wild cattle and driving them to places like New Orleans to sell.[241] But markets were few and far between, and moving cattle to market without expertise was difficult. The settlers' early cattle drives were not particularly profitable; their livestock often ended up as food for their own tables.

Soon, however, these settlers learned from their more experienced neighbors, Mexican ranchers and *vaqueros*, how to more efficiently raise and drive cattle. In addition to borrowing the strategies and gear of the *vaqueros*, including lassos and cowboy hats, American cowboys also began to steal horses from Mexican ranches.[242] These horses were descendants of the earliest Spanish horses in North America, and Mexican ranchers had been breeding them to herd cattle for quite a while. Cowboys, however, often found the wild Spanish horses running free in Texas to have the best herding instincts, even better than the animals specifically bred for the job.

The cowboys bred the rustled Mexican horses with

captured wild horses to create the foundation stock for the Texas cow pony, eventually transforming the San Antonio region into one of the world's biggest horse markets.[243] (In the mid-1800s, the Quarter Horse and other breeds, along with the American bigger-is-better mentality, would make their way West, contributing to the sizing-up of the cow pony.)

Texas cowboys were some of the first nonnative people to successfully make use of the vast, dry Great Plains. [244] They did so by adopting a nomadic lifestyle, like that of the Horse Nations following buffalo herds. But instead of following buffalo to hunt them, the cowboys herded cattle across the plains to graze their way to market. These early cowboys were sensitive to the fragility of the grasslands and moved the cattle along steadily to avoid overgrazing.

It took a lot of horses to herd cattle, and these horses became not only a cowboy's work partners but his companions on the range. Each cowboy required a string of at least ten horses to avoid wearing any of them out. Horses specialized in different tasks.[245] For example, some were skilled in swimming cattle across rivers, some in roping — holding a line tight once a steer had been roped, then backing and sidestepping to keep it from charging — and some in cutting — separating individual livestock from the herd.[246]

A wrangler was in charge of the cow pony herd, which was called a *remuda*.[247] Every morning, he would round up a horse for each cowboy from the *remuda*. Some horses were barely broken, whether fresh from the wild or domestic but not properly trained. They would buck and try to throw off their riders. Cowboys had to be excellent riders to do their jobs, which included learning to control difficult or unbroken horses.

Early cowboys had an interesting relationship with wild horses. Over the years, they captured thousands upon thousands of them to train as cow ponies. Wild horses also caused cowboys headaches, stampeding their cattle and stealing cow ponies to join their wild bands.[248] Nonetheless, these cowboys had a respect for wild horses as stunning symbols of freedom and strength and life on the open range. They required daring to capture, but they became good friends and work partners for the cowboy who succeeded. As J. Frank Dobie, author of the seminal work *The Mustangs*, wrote, "To rope and ride a 'desert horse' was an achievement beyond the pale of commonplace living."[249] These early cowboys coexisted with wild horses and were not bent on destroying them.

Sometimes cowboys would gather around for entertainment and watch as their buddies tried to master their unbroken horses, which they called broncos. Soon, this practice evolved into informal bronco riding contests to entertain onlookers in the towns and stations where the cowboys delivered their cattle—the origins of the rodeo.

Despite some long cattle drives to Illinois, Colorado mining camps, and New Orleans—to sell to the Confederacy during the Civil War—the cattle industry remained small and rather unprofitable until the last half of the 1800s.[250] The arrival of the transcontinental railroad would change the cattle industry and just about everything else on the Great Plains. We'll explore this transformation in a few chapters.

20

TEXAS RANGERS AND THE COMANCHE

A s soon as European settlers arrived in North America, they began to make agreements and treaties to remove native peoples from the land they wanted for themselves. A cycle of peace, followed by violence, followed by retaliation often arose. But little by little, the whites whittled out larger and larger swaths of land for settlement.

Then a series of laws passed throughout the 1800s formally authorized the U.S. government to forcibly remove indigenous peoples from their lands. President Andrew Jackson believed that white settlers and Native Americans could not live together peacefully, so he successfully pushed for legislation requiring the permanent relocation of Native Americans.[251]

An 1834 law established that all U.S. territory west of the Mississippi, excluding Missouri, Louisiana, and Arkansas, would become "Indian country."[252] No whites could settle in Indian country, also called Indian territory, nor could they trade there without permission.

In reality, though, white settlers were streaming west to claim new lands and did not want to give up the majority of the country west of the Mississippi to Native Americans. So the law was immediately revised to push the boundary of Indian country further west and shave down its size.[253]

When the California Gold Rush began in 1848, cascades of Americans barreled across the country in search of fortune and settled where they pleased, ignoring the bounds of Indian country entirely. Then the Indian Appropriations Act of 1851 designated funds for the government to move Native Americans of the West onto reservations in Oklahoma, officially shrinking Indian country once again.

The American people rationalized these policies with the notion of manifest destiny.[254] Journalist John L. O'Sullivan had coined the term in 1845. It was his second use of it in a widely read column that influenced the nation: "And that claim is by the right of our manifest destiny to overspread and to possess the whole of the continent which Providence has given us for the development of the great experiment of liberty and federated self-government entrusted to us."[255] This idea that whites had the God-given right and destiny to spread across North America, dominating the land and its people by propagating democracy and capitalism resonated with many Americans during the period of westward expansion.

Removing Native Americans from valuable lands in the East was less complicated than removing them from the West. Indigenous peoples of the East were largely agricultural, so they had permanent dwellings that made them easy to find. And more significantly, they did not have mounted warriors, whose agility and battle skills would have given them an advantage over the early weaponry of the U.S. government. Many native peoples of the East owned horses by this point, but they did not fight on horseback. U.S. forces had more powerful weapons and were able to subdue the indigenous peoples of the East using their traditional military strategies, as well as political pressure.

Everything was different in the West. The Horse Nations, especially the Comanche, were elusive and had the upper hand over traditional weapons and strategies of the U.S., striking and then disappearing into the immense grasslands.

The first life-changing assault on the Comanche came not from the government but from diseases brought by European settlers. Through the 1830s and 1840s, smallpox and other diseases ravaged the Comanche population, and many other native populations in contact with white settlers. In 1849, cholera devastated the Horse Nations as wagon trains crossed the country to California in search of gold, spreading disease across the plains and wiping out entire bands of native peoples.[256] Within a decade or two, the Comanche population plummeted from its peak of about twenty thousand to around four thousand.

Texas had been trying and failing to fight off the Comanche long before it became a U.S. state. During Texas's short tenure as an independent republic from 1836 to 1845, its government adopted an intolerant stance against native peoples; Texas wanted to exterminate them, or at least push them out of the territory.[257]

During this time, Texas organized the famed Texas Rangers to patrol borders and seek retribution against Native Americans, especially the Comanche, who endlessly raided white settlements that bordered and encroached upon Comancheria. As we know, Mexico had welcomed American settlement in Texas as a buffer against Comanche attacks. The early Texas Rangers had little success in their counterstrikes again the Comanche; they were no match against Comanche battle skills.

These first Rangers were young and ill-equipped to fight mounted warriors.[258] They rode whatever horses they could find. They had to dismount to discharge their one-shot Kentucky rifles, which meant they did not stand a chance against mounted warriors who could shoot twenty arrows from horseback in the time it took a Ranger to shoot and reload once. And they were almost defenseless against Comanche warriors wielding the deadly fourteen-foot lances they used to kill buffalo at a full gallop. Each year, around half the Rangers were killed.

But in 1840, John Hays stepped in to lead the San Antonio

Rangers, transforming their abilities by teaching them to mimic Comanche fighting techniques.[259] Quality of horses was essential; Hays only hired men who had strong, swift mounts. These Hays Rangers became nomads of the plains, like their targets. They learned to shoot their rifles and single-shot pistols while charging on horseback, something no other white Americans were doing at the time. They also employed a favorite Comanche technique in their attacks: the element of surprise. Eventually they began to have some successful encounters against the Comanche, though their weapons were still a weak match for the swift, practiced, mounted Comanche warriors.

Then a man named Samuel Colt invented a revolving pistol. Instead of carrying a single bullet, this first revolver carried five.[260] Initially, no one in North America seemed interested in such a weapon. They had no use for it. But no one else was trying to fight off mounted Native American warriors with superior battle skills.

Hays had figured out that having hardy, agile horses was one piece of the equation to conquering the Comanche, but he also knew the Rangers needed better weaponry. By 1843, they were carrying Colt revolvers and shooting them from the saddle. The revolvers were not perfect, but they catapulted the Rangers forward in their capacity to fight mounted warriors. They began to see more effectiveness in their engagements with the Comanche. (Colt's designs had been so unpopular with the rest of America that he slid into poverty for five years. Eventually word spread about the efficacy of the weapon in Texas against the Comanche and during the Mexican War, and he started to produce it again, this time as a six-shooter.)

By emulating the Comanche, Hays had begun to crack the code to defeating them. But he left Texas for California in 1849 in search of gold, and his knowledge seemed to have disappeared along with him.[261]

In the 1850s, following the Mexican War and the acquisition of Texas, the U.S. government assumed the role of trying to subdue Native Americans in the region. Instead of learning from Hays's successes, the army sent laughably quaint "elite" fighters, known as the dragoons, up against the Horse Nations.[262] They wore fancy uniforms, rode heavy horses that lacked speed and stamina, and dismounted to fight with their single-shot pistols and swords.

Predictably, they were an utter failure. Unable to subdue the Comanche, the U.S. maintained a policy of defense rather than offense against them for most of the 1850s,[263] leaving the Comanche virtually unchecked and in command of the southern Great Plains.

However, in 1855, about four hundred Comanche agreed to move to a small reservation in Texas after they had been ravaged by disease and starvation following the white settlers' slaughter of the food sources upon which they depended.[264] The government's goal was to make farmers out of the reservation Native Americans, to get them to stay put and stop traipsing across the plains to hunt buffalo, raiding along the way. But the Comanche were not interested in farming, and soon skirmishes between the Comanche and the white farmers whose land surrounded the reservation intensified to a dangerous level. So the government sent this small group of Comanche packing to another reservation in the Indian territory of Oklahoma.

Texans grew tired of the constant raids from the remaining Comanche. In 1858, the state reorganized the Rangers—relearning that they needed to hire skilled riders, find horses with good stamina, wield six-shooters, and emulate Comanche battle techniques—and went on an offensive against the Comanche, following them into Comancheria. [265] The Rangers' successful attack, known as the Battle of Antelope Hill, showed that it was possible to defeat the Comanche with the right techniques and sprang the U.S. Army from its defensive stance. To demonstrate its new confidence and offensive strategy, the army then went out and massacred seventy unsuspecting Comanche north of the Red River.

Suddenly, the tide seemed to turn against the fierce Horse Nation. Between the substantial reduction in the Comanche population from disease and the U.S. military recalling how to use some of the mounted warriors' own techniques against them, it seemed like the Comanche were on the brink of collapse.[266] Except that in the East, another tide was turning. War was brewing, and it would soon drain the West of the men fighting the Comanche before they had a chance to defeat them.

21

THE PONY EXPRESS

B y the middle of the 1800s, westward expansion had pushed the border of the U.S. all the way to the Pacific Ocean. Around four hundred thousand people had traveled the Oregon Trail to begin new lives in the Pacific Northwest. Thanks to the Gold Rush, California had a brand-new, nonnative population of 380,000. Isolated mining towns sprang up throughout the mountain West. Ranches filled Texas. The philosophy of manifest destiny thrived.

Before telegraphs, telephones, the transcontinental railroad, or automobiles, the vast country needed a way to communicate from coast to coast. In 1858, Butterfield Overland Mail Company made one of the first attempts to link East and West by a stagecoach service between St. Louis, Missouri, and San Francisco, California.[267] The stagecoach carried mail and passengers along a grueling twenty-eight-hundred-mile route, which took around twenty-five days driving twenty-four hours per day at five

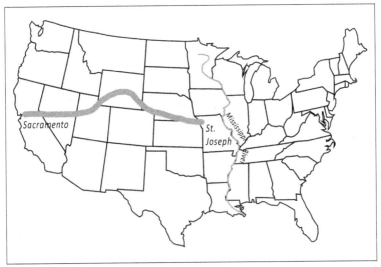

Approximate Pony Express route between St. Joseph, Missouri, and Sacramento, California, 1860-1861. (Some western states were still territories at the time.)

miles per hour, with continual, rapid stops for fresh horses.

The stagecoach required larger horses than the compact mustangs of the West, so Butterfield bought a thousand heavier horses and five hundred mules to distribute among the stations along the route.[268] Each station kept a pasture of horses to refresh the stagecoach, but these animals, especially attractive since they were larger than most horses in the region, often fell prey to horse rustling by white settlers and Native Americans.

Although it was a start, the country needed a faster way to communicate than the stagecoach could offer. The answer? Mail delivered by horseback. In April 1860, the Pony Express opened for business between St. Joseph,

Missouri, and Sacramento, California. The route was nineteen hundred miles and had around 190 stations for riders to switch horses.[269] Four hundred twenty fast horses with incredible stamina—most with Spanish blood, including many California mustangs—were divided among stations.[270]

Riders were lightweight and carried little gear. Saddles and bridles were also light. They carried the mail in locked leather pouches called *mochilas*. It initially cost an exorbitant $5 per half-ounce to mail a letter,[271] or approximately $145 in today's dollars. The price dropped to a still-steep $1 per half-ounce by the end of the operation.

A rider rode about seventy-five miles per shift, stopping at a station every ten to fifteen miles for a fresh horse. Transfers took between two seconds and two minutes and involved switching the *mochila* to the new horse and grabbing a piece of bread or cup of coffee.[272] Once a rider completed a seventy-five-mile stretch, he slept at a station until the return mail arrived. Then his next shift began.

The Pony Express only lasted for eighteen months until October 1861, when the first telegraph line between East and West was completed. Now news could travel across the country in a matter of minutes. Despite its short tenure, the Pony Express joined the U.S. at a crucial time in its history: the lead-up to and outbreak of the Civil War. Riders even brought word of Lincoln's presidential election, as well as his inaugural address.[273] The romantic notion of horses galloping across the majestic plains and the mountain West to deliver the mail became almost as enduring an American image as the western cowboy.

PART V

THE EAST
EARLY TO MID-1800s

22

RESHAPING THE LANDSCAPE OF A NATION

P rior to the Revolutionary War, the horse was essential to life in some areas of North America, but less so in others. Early Spanish explorers depended heavily on horses in the West, both to transport them through the immense, arid region, and to dominate the native peoples they encountered. That is, until those native peoples gained access to horses and flipped the script, using the animals to overtake the Great Plains. In the East, thick forests and swamps limited the effectiveness of the horse, but early settlers used the animal to varying degrees to travel and to clear land for agriculture.

But during the 1800s, the horse transformed into the primary energy source of the century. In doing so, it actually reshaped the landscape of the new nation.[274] Or, more accurately, humans reshaped the landscape so the horse could more efficiently fuel the developing nation.

While Lewis and Clark were exploring the western lands

the U.S. had obtained from the Louisiana Purchase in 1803, Americans east of the Mississippi River understood it was necessary to better link the states and territories. Citizens needed to be able to communicate with each other, and as they moved further west and away from waterways, they needed overland ways to transport and sell goods. The country's first president, George Washington, was also concerned that far-flung, isolated citizens might lose their loyalty to the new nation and try to found their own countries on U.S. soil.

These needs for connection spurred a transportation revolution that would include the construction of roads, canals, and railroads, especially across the Northeast, all of which required horse power to build and operate.[275]

The militaries fighting the Revolutionary War had carved out new routes through and between the states, and the new country went to work constructing passable roads from those routes and others to truly unite the states.

During this time, only three people in the country were considered experts in building artificial roads that were more efficient than worn, dirt tracks, and even they did not agree upon a method.[276] Their styles involved either varying layers of stone or wooden planks. Regardless of the materials the builders chose, the goal was always the same: to reduce the draft for a horse, meaning to decrease the resistance against the horse as it pulled, thereby increasing its energy efficiency.

Citizens wanted roads, but they grumbled about paying

for them through taxation.[277] Toll roads, known as turnpikes, were one solution to cover the cost. After the 1795 opening of the Philadelphia and Lancaster Turnpike, which was widely traveled by the Conestoga horses hauling their Conestoga wagons from farms into the city, turnpikes sprouted all over the Northeast.

While the South lagged far behind in road building, the North made useful road connections between and within states, opening up travel and trade to riders on horseback and horse-drawn wagons hauling goods to sell to new markets.

Stagecoach lines also began to appear as an early form of mass transportation, some operating between local towns and others making long-distance trips connecting major cities.[278] These stagecoaches, drawn by four to six horses, had to stop every fifteen to twenty miles for fresh horses. Fashioned by the requirements of horses, communities emerged around these stations with inns for travelers and drivers, stables for horses, farriers (people who care for horses' hooves and fit them with shoes), and farmers who grew hay and oats to feed the horses.

Stagecoach companies also received contracts from the U.S. Postal Service to deliver mail, which accounted for as much as a third of their income.[279] Businesses arose to make a variety of new vehicles for the expanded traffic.

Canals became another important way to transport goods inland. The Erie Canal, built between 1817 and 1825, connected the Great Lakes to New York City and the

Atlantic Ocean, opening up a significant new trade route. It was the largest public works project attempted in the world at that time and an incredible feat of engineering.[280] The canal, along with the others that fully linked the Great Lakes, would also have enormous environmental implications by catalyzing the incursion of invasive species, which have been devastating these essential fresh water ecosystems for two centuries now.[281]

The Erie Canal, which required at least two thousand horses to dig and construct, was built so that horses and mules could tow barges filled with goods through the waterways.[282] Long ropes connected the animals to the boats, and they walked up the canal on a path along the bank to move them. Towing a boat in water is an extremely efficient way to move freight. Two packhorses can carry a total of about four hundred pounds on their bodies. Two horses harnessed to a wagon can pull about a ton. Two horses towing a barge can haul fifty to eighty tons.[283] By the middle of the 1800s, there were eight thousand horses working the Erie Canal.[284] Even once steam engines were invented, they were not used much on the canal, because they caused problematic wakes that eroded the banks.[285]

Around the same time, horse ferries came into fashion, used for transporting people across rivers, lakes, and sounds. Horses would walk on a treadmill on the boat; the treadmill turned a waterwheel to power the boat.[286] The first commercial horse ferry sailed between Brooklyn and Manhattan. It used six to eight horses and made the crossing

in fewer than twenty minutes.

Horse ferries subsequently sprang up along the East Coast and well inland. They were popular on Lake Champlain to connect New York and Vermont. In 1983, the wreck of the Burlington Bay Horse Ferry was discovered in the lake, fifty feet underwater, where it is still accessible to intrepid divers.[287] The last documented horse ferry in the country operated until 1920 on Tennessee's Cumberland River.

It may not seem intuitive, but the development of the railroad drastically increased the use of horses in the country in the middle of the 1800s.[288] Between 1840 and 1850, the horse population increased by 12 percent. But between 1850 and 1860, when the railroad was rapidly expanding, the horse population grew by 51 percent. The states with the most miles of railroad track had the most horses.

Horses were essential to railroad operations. Not only did they do the heavy work associated with constructing the railroads, they also pulled the first trains before steam engines were well established.[289] In addition, they hauled all the freight and passengers to and from stations.

Railroads were not public enterprises; they were companies looking to turn profits. Early on, they did not cooperate with each other, there was no standardization of equipment, and they all had different stations, even within the same cities.[290] (Steam engines, at risk for exploding and catching a city on fire, were often relegated to the outskirts.)

Railroads were useful for long-distance hauls, but since they ran on a set track, they lacked flexibility. So people and freight making connections required horse transportation between depots. Like the stagecoach communities, new businesses, such as livery stables, stagecoach lines, inns, and restaurants, planted themselves in these areas.

The greater need for horses to haul goods and passengers along the expanded transportation network increased demand for heavy horses. Breeders had been raising draft horses since the Revolutionary War, but they responded to the demand for even larger animals for industrial work by importing draft horses from Europe.

French Percherons arrived in the U.S. in 1839 and rapidly became the most popular draft breed for agricultural work.[291] Percheron stallions were often bred with local mares to make their size more manageable.[292] Belgians, Suffolk Punches, and Cleveland Bays also became popular on farms. Massive, slow, strong Shires and Clydesdales were used almost exclusively for the heaviest hauling in cities.[293]

These draft horses fulfilled the increased demands of hauling throughout most of the 1800s. But during this time, American desire expanded not only for drafters, but for all sorts of specialized horses to power the country's growing industries.

The transportation revolution of the 1800s, which centered around reshaping the landscape to most efficiently harness the power of the horse, connected the nation by improving communication, easing travel, and opening up far-flung markets for selling goods.

23

EMERGING AMERICAN BREEDS

D rafters were not the only horses in demand during the 1800s. Throughout the century, several of the first truly American breeds emerged to meet the requirements and desires of different communities, whether for farm work, riding, driving, or entertainment. At first there were more regional types of horses than set breeds. As the century progressed, however, breeders became more interested in refining these regional horses into particular breeds with somewhat fixed characteristics, as defined by breed associations. These breed associations began "stud books" to carefully document lineage.

I n the late 1700s and early 1800s, New Englanders continued to clear their heavily wooded lands to develop farms. They did not have a lot of money and needed a versatile animal to serve as a saddle horse, harness horse (carriage puller), stump puller, work horse on the farm, and a casual race horse when fair day or other entertainment opportunities arose. Such an all-around horse was difficult, or nearly impossible, to come by, especially as the Narraganset Pacer was disappearing after heavy exportation to the West Indies sugar plantations.

In the late 1780s or early 1790s, an unusual foal was born in Massachusetts. More than 150 years later, Marguerite Henry turned him into a modern household name when she published a fictionalized account of his life, *Justin Morgan had a Horse*. Even today, the story of Justin Morgan is well known.

The evidence we have about the horse's actual life is slim but still makes for a great narrative. He was a small bay (dark brown with black points) that grew to around 14 hands and 950 pounds. His breed origins were unknown.[294] Around the age of three, he came under the ownership of a music teacher named Justin Morgan, who lived in Vermont. Over time, it became apparent that the horse, whom Morgan called Figure, was an incredible and versatile athlete. Despite his small stature and the criticism he received for it, Figure won log-pulling contests, beating out draft horses and entire teams of horses. He won walking races. He won trotting races pulling a carriage. He won running races.[295]

After Justin Morgan died, people began to refer to Figure as Justin Morgan or The Morgan Horse. He was incredibly prepotent, meaning his offspring inherited many obvious characteristics from him, and he went on to become the first individual horse to sire an entire breed: the Morgan. The breed, at 14.1-15.2 hands, would become one of the most popular horses in American history due to its versatility as a saddle horse; harness horse; ranch horse; a choice of 49ers headed to the California Gold Rush; and a favorite of the U.S. Army during and after the Civil War, used for, among other purposes, trying to subdue Native Americans of the West.

By the mid- to late 1800s, you could find Morgans in every state and territory of the U.S.[296] However, as the century drew to a close, trends changed. Americans began to demand bigger horses with longer legs. They combined Morgans with larger horses to create more fashionable animals, and nearly bred the Morgan out of existence. Luckily, some areas of Vermont still used the original Morgan horse type, and the breed that exists today was preserved through those animals.

A t the same time Justin Morgan was turning heads, harness racing—pulling a carriage while trotting or pacing—was becoming the rage in New England. (See chart on page 93 for a detailed description of gaits.) One reason for the popularity of this type of racing was that Puritans considered running races on horseback sinful but found trotting and pacing while pulling a carriage acceptable.

To create fast harness racers, New Englanders crossed their horses with Narragansett Pacer blood with Thoroughbreds (and, in particular, with a famous

Thoroughbred named Messenger and his offspring, Rysdick's Hambletonian), as well as Morgans (and, in particular, a famous trotter named Ethan Allen).[297]

The result of these combinations became known as the Standardbred. The first stud book of the Standardbred required that for a horse to be registered in the breed, it must trot a mile in at least two minutes thirty seconds or pace a mile in at least two minutes twenty-five seconds.[298] Because the horse had to trot or pace to a standard time, it was called a Standardbred. Today, due to improvements in breeding, equipment, tracks, and training, Standardbreds are required to be faster.[299]

I n the South, the wealth of white plantation owners continued to grow throughout the 1700s and into the 1800s, and with it, a desire for well-bred, fashionable horses. Following George Washington's promotion of the mule, large plantations in Virginia and the Carolinas had taken to using mules for heavy field work. But plantation owners and their families wanted smooth, handsome saddle horses for their own transportation as they managed field work or rode to town. They mixed their Spanish horses, the gaited ones, with horses that had Narragansett Pacer blood to try to create their smooth saddle horse.

It was this type of gaited horse the explorer and folk hero Daniel Boone and his compatriots brought west in 1775 to found the first white settlement in Kentucky.[300] Over the years, Kentucky breeders refined this horse, grazing it on

their nutrient-rich bluegrass and adding Thoroughbred (courtesy of Messenger offspring), Canadian Pacer, and likely Morgan blood to create a sleek, long-necked, three- or five-gaited animal known as the Kentucky Saddler, and later, as the American Saddlebred. This horse became popular among affluent southerners.

During the Civil War, both sides used the Saddlebred, known for its endurance and steadiness at 15-17 hands, as officer mounts.[301] Generals Grant, Jackson, Lee, and Sherman all rode Saddlebreds. Founded in 1891, the American Saddlebred Horse Association is the oldest horse breed association in the country.

U nlike in Kentucky, where many farms were large and used slaves and mules to work the land, farms in Tennessee were often smaller operations, and the farm owners and their sons worked their land themselves with their horses.[302] They had no use for and could not afford fancy, smooth, artificially gaited animals. They wanted all-purpose horses that were strong but calm for farm work, gentle for the women of the family to ride, attractive enough to measure up to societal norms, and also able to serve as harness horses to pull the family carriage to church on Sunday.[303]

Like early settlers of Kentucky, early settlers of Tennessee also brought their saddle horses from Virginia and the Carolinas. But what they sought through selective breeding was more of an all-purpose horse than the eventual Saddlebred. They combined their saddlers with numerous types: fresh Spanish mustangs from Texas (via the trading center of Natchez, Mississippi), Thoroughbreds, Standardbreds, and Morgans.[304]

Over time, these breeders produced the fine, 15- to 16-hand Tennessee Walking Horse, which naturally performs three ambling gaits that most horses do not—perfect for the rigors of small-farm life in Tennessee.

A s a result of the disruption to horse breeding and importation during the Revolutionary War, as well as the migration of settlers westward toward the new frontier, Virginia's status as a horse breeding center waned during the decades following the war. However, the horse still played an important cultural role in Virginia as an entertainer and a symbol of wealth and status.

The racing that did remain in Virginia began to shift from the short straightaways that Quarter Horses dominated to long, oval tracks that Thoroughbreds did. By the middle of the 1800s, Quarter Horses had fallen out of fashion in Virginia and other racing states.[305]

But as settlers continued to surge westward in search of gold or later to claim their land following the Homestead Act of 1862, which gave away approximately 160 acres of public land to qualified applicants, they brought the all-purpose Quarter Horse with them. As more Americans began lives as cattle ranchers out West, especially in Texas, they realized their Quarter Horses had excellent cow sense, thanks to their Spanish and Galloway origins. Over time, demand for larger horses, as well as changes in the duties of ranch workers, led ranchers to breed up their Spanish cow ponies using Quarter Horses and eventually Thoroughbreds.

The transportation revolution that occurred throughout the late 1800s, especially in the North, simultaneously opened up new horse markets, transforming horse breeding from a regional enterprise into a national one. By the end of the century, a large variety of horse breeds with somewhat standardized characteristics existed throughout the U.S.: all-around work horses, handsome saddle horses, horses with artificial gaits for show, cow ponies, and heavy drafters. These horses, all one species but now heavily differentiated in their characteristics, would power the country throughout the 1800s and into the 1900s.

24

THE TRAIL OF TEARS

Since Europeans arrived in the Americas, they had been working—through cajoling, bribery, treaties, and force—to move Native Americans off their ancestral lands to claim those lands for themselves. The process continued after the colonies declared independence and became the U.S. When native peoples did remain near white settlements, the government's goal was to convince them to assimilate to European-American culture by converting to Christianity, learning English, and settling peacefully on farms instead of practicing a nomadic lifestyle.

In the Southeast, the Native Americans that acquired, bred, and spread Spanish horses east of the Mississippi— the Cherokee, Choctaw, Chickasaw, Creek, and Seminole— were willing to adopt these European-American practices to some degree, so the government left them alone for a while.

But as cotton and tobacco plantations became successful,

southerners clambered for more farm land.[306] The discovery of gold on Cherokee territory in Georgia in 1829 brought the white desire for this land to a fever pitch; the territories of the five indigenous nations were just too valuable to pass up.[307] In 1830, the Indian Removal Act, signed by President Andrew Jackson, allowed the federal government to remove these five peoples from their lands and send them to live in Indian territory west of the Mississippi, which is now Oklahoma.

So began the Trail of Tears, which was not one incident or one westward trail, but rather twenty years of ejecting Native Americans from their Southeast homelands and forcing them west. In numerous waves, these indigenous peoples made their way to Indian territory in Oklahoma by boat, which took several weeks, or by land, which took several months.[308]

The Cherokee were the last of these nations to move west. In 1835, a few Cherokee chiefs finally resigned to government pressure and signed the Treaty of New Echota, agreeing that the nation would move to Indian territory in exchange for $5 million.[309] An uproar followed; these chiefs did not represent the majority of the Cherokee. Around sixteen thousand Cherokee signed a petition to nullify the treaty (although there were actually fewer than that number of Cherokee in the East, so the number of signatures remains suspect), but Congress ignored their demands and ratified the treaty.[310]

By 1838, however, only about two thousand Cherokee

had gone west.[311] Sick of waiting for resolution to the "Indian problem," President Martin Van Buren sent seven thousand troops south to forcibly remove the remaining Cherokee population. They rounded up the Cherokee, and the blacks the Cherokee had enslaved, and detained them in camps before marching them twelve hundred miles to Indian territory. Part-way through the relocation process, the government agreed to allow the Cherokee to manage the relocation themselves, but the results were no less disastrous.

In the camps and along the Trail of Tears, an estimated four thousand Cherokee died of exposure, starvation, and mostly disease—cholera, dysentery, typhus, whooping cough, malaria, tuberculosis. (Around six hundred Cherokee evaded capture and remained on private lands in western North Carolina, establishing the Eastern Band of the Cherokee Nation. The band still lives there today.) Several thousand more Chickasaw, Choctaws, Creeks, and Seminoles died as they rode their Indian ponies or walked the forced journey west.

These Native Americans brought with them their herds of Spanish horses, which had mixed with other breeds by then. Thousands of these horses died along the trail, as well.[312] Times had changed since they had served as the foundation stock for the horses of the first European settlers, often thanks to their distribution throughout the East by these same indigenous peoples. Many Americans now looked at these horses with disdain, considering them scrubby and weak compared to newer, larger, more refined American breeds.

The U.S. government also thought little of the horses of Native Americans, viewing them as obstacles to the goal of convincing indigenous peoples to settle on their western reservations and become productive farmers. The government recognized that killing the horses would break the spirits of the native peoples, so they often found reasons to slaughter these horses or remove them from their owners.[313]

The Trail of Tears represents the demoralizing era of "Indian removal" from the East, particularly of the five nations of indigenous peoples who first spread Spanish horses through the colonies. It was these horses that had allowed early settlers to plow their fields, cut roads, develop horse breeds specialized to their needs and desires—to build a nation. Now they and their owners had been relegated west to undesirable lands.

25

THE CIVIL WAR

E ach carrying nothing more than a rifle, one hundred rounds of ammunition, a saber, and five days of food rations, the seventeen hundred soldiers rode hard for sixteen days straight, covering six hundred miles through the heart of Mississippi.[314] Their primary mission was to create a diversion, so the Union (the North) could ferry troops across the Mississippi River to take the Confederate (southern) stronghold of Vicksburg. Their diversion was to involve destroying railroads and cutting telegraph wires to disrupt supply lines and communications in and out of Vicksburg.

This Civil War cavalry raid of 1863 was one of the most daring and dramatic of the war and demonstrated one of the most effective uses of cavalry: guerrilla-style combat. Colonel Benjamin Henry Grierson and his men damaged fifty to sixty miles of essential railroad and telegraph lines, took five hundred prisoners, seized and blew up three thousand stands of weaponry (a stand is one musket and its accessories), confiscated a thousand horses and mules, and destroyed large quantities of army supplies. They also sufficiently distracted the Confederate army, allowing the Union to transport troops across the river in a move that would help them capture Vicksburg in the months to come—their primary mission and a massive strategic win for the Union and for the morale of Union citizens.

Unlike Civil War images we often see in textbooks of European-style cavalry charges across open battlefields, that type of activity played only a small part in the war. Horses did, however, play enormous, essential roles. Scouts and officers rode horses, and cavalry often used them in skirmishes and guerilla attacks, though rarely for open charges. But far and away, the majority of horses hauled supplies, including artillery, equipment, and food.[315]

During the war, the Confederacy and the Union fought to establish whether slavery would remain the kingpin of the South's agricultural economy. The Civil War was the first industrialized war in the U.S., meaning the armies mass-produced weapons and supplies, and railroads transported large volumes of troops and goods. This industrialization made the war enormous in scale compared to previous American wars. Since railroads could only bring supplies to a limited number of destinations, horses powered almost every aspect of the war.[316]

In fact, a war of that scale during that time period could not have happened without a massive supply of horses.[317]

Luckily for both sides, they were well-stocked at the beginning, because breeders had already been pumping out horses to keep up with the industrial demands of the growing nation.[318] At the start of the war, there were about 3.4 million horses in the North, 1.7 million in the South, and eight hundred thousand in the border states of Kentucky and Missouri.[319]

Both armies were too large to live off the land; they could not gather enough food and supplies locally while they moved around to sustain themselves, as militaries had during previous wars.[320] Instead, horses had to haul supplies, including food for the soldiers and forage for themselves, in wagon trains along with the troops. These wagon trains sometimes stretched an astounding eight miles long, comprised of approximately three thousand to six thousand horses and mules pulling 850 wagons.[321]

During the first two years of the war, the Confederacy had a clear cavalry advantage over the Union. Although the Confederacy had fewer horses overall, since many plantations used mules instead of horses for heavy work, many southern men were expert horseback riders. They still did not have a well-developed system of roads in the South, so they needed to ride for transportation, as opposed to being able to hitch rides as passengers in carriages and wagons, like in the North. The South still had a distinct horse culture, and horses and horsemanship were also very much a status symbol, especially among the wealthy. When southern men enlisted, they were required to bring their own horses with them, so they had the advantage of familiarity with the animals they rode, as well.[322] They also employed their horses strategically for guerilla battles for which the Union was initially unprepared.[323]

The Union had more horses, especially of the draft variety for the heavily industrialized work happening in the region at the time. But Northerners were not the horsemen of the South.[324] Despite ongoing skirmishes with the Horse Nations in the West, the Union had learned little about the advantages of light cavalry horses and instead preferred their drafters, perhaps envisioning they would be involved mostly in European-style open charges that favored them.[325]

Horses had a tragic go of the Civil War. By the end of it, well over a million of them were dead.[326] In comparison, between 620,000 and 850,000 soldiers died in the war.[327] Horses died from wounds in battle, from being overworked; ill cared for; starved; and from illnesses, such as glanders, a highly contagious and fatal disease in its acute form that causes pulmonary or bloodstream infections, as well as ulceration of the skin, lymph nodes, and other organs.[328] Horses can also transmit the disease to humans. One glanders outbreak in 1864 killed eleven thousand horses within a few months.

During the war, the average lifespan of an artillery horse, which pulled cannons onto the battlefield, was seven and a half months.[329] Dead horses littered battlefields and countrysides. The logistics of supplying and constantly resupplying armies with enough horses to support the war, along with the horses' forage, were mind-boggling.

Horses traveled long, grueling distances during the war and were in constant need of re-shoeing to keep their hooves healthy and functional. One small innovation in this area made a huge impact on the war: the 1857 invention of a machine that mass-produced horseshoes.[330] If horseshoes had had to be handmade during the Civil War, as they always had been before 1857, it is not a stretch to say the war could not have occurred on the magnitude it did.[331] There was no way armies could have handmade enough horseshoes in the field to keep their horses functioning.

As the years progressed, the supply of horses in the South became scarce, since they had fewer to start, while the North continued to hold an adequate supply, in part by confiscating horses from southern breeding centers as they began to take control of the region.[332] In addition, Union soldiers had spent the early years of the war improving their riding skills to the point that they finally made formidable foes on horseback.

Eventually, southern soldiers had to start taking leave from the war to rustle up new horses to ride once their own had died.[333] Decreasing availability of horses in the South meant a decreasing ability to fight. Without horses, there could be no movement of troops, no positioning of cannons, no scouting, no guerilla raids.

The dwindling horse population of the Confederacy was indicative of the army's generally shrinking resources, as well as the logistical challenges both armies had in keeping enough horses alive and fed to power the war. When General Robert E. Lee surrendered his Confederate army to General Ulysses S. Grant in Appomattox Court House, Virginia, he asked Grant to allow his soldiers to keep their saddle horses, knowing it would be nearly impossible for them to find horses to work their farms when they returned home.[334]

During this meeting, referred to as The Gentleman's Agreement due to both parties' civility, Grant obliged to this unusual request in the spirit of national healing. He wrote in the surrender documents: "The arms, artillery, and public property are to be parked and stacked and turned over to the officer appointed by me to receive them. This will not embrace the side arms of the officers, nor their private horses or baggage. This done officers and man will be allowed to return to their homes not to be disturbed by United States authority so long as they observe their parole and the laws in force where they may reside."[335]

Lee also asked for forage to feed the remaining, starving Confederate horses under his command.[336] But Grant had none to give.

As the first industrialized war in the U.S., the Civil War relied heavily on horses as movers of massive numbers of troops and quantities of supplies, mounts for officers and scouts, and as cavalry used to stage guerilla attacks. This war could not have been fought without significant horse power and mass-produced horseshoes. Millions of horses contributed to the war effort on both sides, and more than a million lost their lives in the process.

While most Civil War battles took place in the East, the war had a major influence on dynamics in the West; men abandoned the West to join both sides of the war, leaving the frontier largely abandoned by those Americans trying to keep the Horse Nations in check. Following the war, however, the U.S. Army would turn its attention toward taking control of the West once and for all.

PART VI

THE WEST
MID- TO LATE 1800s

26

WESTWARD TRANSFORMATION: THE RAILROAD

Near the close of the Civil War, the West was sprinkled with settlements: rancher-cowboys in Texas, who were still not making much profit; mining towns spread across California and the mountain regions, booming and busting; farming communities in the Pacific Northwest; and, thanks to the Homestead Act, industrious new property owners heading west with hearts set on taming unruly lands.

The Homestead Act was part of an American development strategy that differed from Spanish colonization of the Americas. The Spanish first set up missions, run by the church and military. These missions broke ground, attempted to subdue the local indigenous peoples, and offered protection to settlers.

The U.S. government, however, preferred citizens to do the work of settling new territories themselves, offering free land to entice westward migration. Once settlers established themselves, the government sent back-up, including the Buffalo Soldiers—black army regiments assembled after the Civil War to do the grueling work of patrolling the frontier and quelling disputes between settlers and their white or Native American neighbors.[337]

Wild horses supplied the early farms and ranches of the West. All these settlers had to do was figure out how to round up the mustangs in their backyards, and the horses were theirs for the taking. Wild horses were also trailed East to supply farms and cities.[338]

Life presented numerous challenges for the homesteaders who elected to take a chance on the Great Plains. For starters, the skills they used to farm the rich soils of the East did not transfer to these fragile grasslands with little topsoil or rain.[339] Government promises of a new life often ended in crop failure.

Second, although the U.S. claimed ownership of the Great Plains, thanks to the Louisiana Purchase, and had begun removing some Native Americans to reservations, the Horse Nations still dominated this region and did not appreciate encroachment by white settlers. Nothing struck fear in the hearts of these settlers like roving, raiding bands of mounted warriors, and hostile encounters sometimes occurred. [340]

Third, the buffalo, which the Horse Nations followed and on which they depended, caused problems for ranchers on the plains. Besides bringing native peoples near and allowing them to continue their nomadic ways, the buffalo ate grasses settlers wanted for their livestock and trampled their crops. Wild horses did the same.

This was the scene on the Great Plains when one event dramatically transformed the region and its horse culture: the 1866 arrival of the railroad in Kansas. Seemingly overnight, the culture of the Great Plains moved from one of subsistence living by the indigenous peoples and a smattering of cowboys and white settlers to domination and extraction by the rest of the country, as well as foreign investors.[341]

Railroad access to the Great Plains meant they were no longer a destination only for an adventurous soul willing to undertake a long, arduous journey on horseback or by wagon. Now just about anyone could hop a train headed west. And easy in meant easy out; it was now profitable to remove resources from the middle of the country and sell them in large eastern markets. Anything that could be thrown on a train and hauled off to sell was: cattle, horses, timber, minerals.[342]

The railroad opened up large, urban markets that suddenly converted cattle ranching into a profitable business. Overseas and eastern investors snapped up small, cowboy-owned ranches, combining them into massive corporations.[343] They took over much of the Great Plains and Pacific Northwest for grazing land, and they wanted the Native Americans, buffalo, and wild horses out of their pastures. In time, they got what they wanted, at the expense of the people and animals who had co-existed on the plains for centuries. This transformation meant the beginning of the end for the horse culture of the native peoples of the Great Plains. The next three chapters explore this showdown, involving stockmen, mustangers, the U.S. Cavalry, and the Horse Nations.

27

THE COWBOY'S HEYDAY FOLLOWED BY A SWIFT END TO THE OPEN RANGE

The arrival of the transcontinental railroad in Kansas in 1866 catapulted cattle ranching into a profitable industry. Cattle could finally be moved efficiently to large markets, like Chicago. The heyday of cowboy life on the open range had begun, though it would only last about twenty-five years before the abrupt end of the open range.

Texas cowboys, with their strings of Spanish cow ponies, drove cattle great distances across the plains to cattle stations at railway stops. In the early days, they worked to move cattle to fresh grass before they had overgrazed the fragile grasslands.[344] These cowboys also coexisted with wild horses, living alongside them and sometimes capturing them for their own purposes, but not seeking to eradicate them.[345]

Although the western genre of literature, and later film, created the legend of the cowboy in the image of a white man, about one-fourth of western cowboys during the late 1800s were black.[346] Southern whites had enslaved many of these men and brought them to Texas prior to the Civil War to work their ranches. When the white ranchers went east to fight for the Confederacy, it was the men they had enslaved who tended the ranches in their absence.[347]

When the war ended, abolishing slavery, the white ranchers returned home to the West. They realized that to rebuild their operations after the difficult war years, their only choice was to hire the laborers with the most experience managing horses and cattle: the men they had once enslaved.

During a time when the place of free blacks in society was in question, black cowboys of the West experienced a higher status than many other blacks in the South. In fact, many formerly enslaved blacks who had previously worked cattle in the Southeast chose to make their way to Texas to work as cowboys.[348] While the lives of many black cowboys of this era went unrecorded, a few did become treasured legends, such as Nat Love—one of the only black cowboys of the era to leave behind an autobiography, and a dramatic one at that—and Bill Pickett—one of history's most famous rodeo riders, known for his incredible trick of leaping off a galloping horse onto a steer's back, wrestling it to the ground, and biting its lip to shock it into submission.

On the range, black cowboys had a degree of equality and independence that many other blacks did not experience in the years following the Civil War.[349] They also received pay equal to their white colleagues'.[350] In short, their horsemanship and ranching skills elevated their position in society during the cowboy's prime on the western range.

History has also largely left out Native Americans and Mexicans from cowboy lore. Women were banished from much of the historical record of the range, too, except in early dime-novel depictions as prostitutes or helpless maidens in distress. But many women of the West during this era defied such plot lines. Some worked cattle alongside their husbands, some inherited or established and ran their own ranches, some rustled cattle.[351]

Others competed in rodeos, riding bucking broncos and roping steers along with their male compatriots.[352] Annie Shaffer, of Arkansas, and Prairie Rose Henderson, of Wyoming, were some of the earliest female bronco riders.

During this time, rodeos exploded from informal gatherings of cowboys showing off their bronco-riding abilities into formal contests that charged admission, awarded prizes, and included female competitors. These early rodeos were typically part of Fourth of July celebrations.

The term "cowgirl" did not become popular until Theodore Roosevelt uttered it in 1900, in reference to fourteen-year-old Lucille Mulhall, a skilled performer and worker on her family ranch.[353] After watching Mulhall's roping performance in front of a crowd of twenty-five thousand, Roosevelt struck up a friendship with her and her family. The Mulhalls invited Roosevelt to their ranch, where he witnessed more of Lucille's riding and roping. Roosevelt, the vice-presidential candidate at the time, told her if she roped a wolf, he would invite her to his inaugural parade. Within three hours, she returned dragging a dead wolf behind her horse. The following year, Roosevelt made good on his promise, inviting Mulhall and her family to participate in the inaugural parade.

Wild West Shows—the most famous of which was orchestrated by Buffalo Bill Cody, a former soldier who claimed to have been a Pony Express rider and to have slaughtered more than four thousand buffalo in a seventeen-month period—also became popular in the late 1800s. They were often part rodeo, part dramatic portrayal of western stereotypes of cowboys, Native Americans, and life on the frontier. Rodeos and Wild West Shows played into the advent of western literature, creating and solidifying the romantic image of cowboy life on the range for Americans in the East.

Ironically, the arrival of the railroad on the Great Plains allowed for the peak of the long cattle drives to occur and the romantic image of the cowboy to form, but it also catalyzed a radical and abrupt shift in the dynamics of life on the plains.

When absentee investors swooped in and bought up many of the small ranches, rancher-cowboys became employees of these large corporations instead of their own bosses, no longer able to make their own rules to navigate the plains.[354]

The cattle industry outgrew Texas and spread north throughout the Great Plains, spilling into the Pacific Northwest. But investors saw barriers to full dominance of the plains: the presence of buffalo and the nomadic Horse Nations that followed them, as well as the presence of wild horses eating up the grasses they wanted for their cattle and sheep. They did not want the horses' natural predators, wolves and mountain lions, in their pastures eating their livestock either. These stockmen wanted to keep people and animals out of what they viewed as their giant pasture. In a matter of years, the wild horses the cowboys and early settlers respected and used for their own purposes turned from symbols of liberty and history warranting respect and awe into nothing but varmints.

Over the decades that followed the arrival of the railroad, the U.S. government, now finished with the Civil War, stepped up its strategy to rid the West of native peoples so settlers, miners, and stockmen could claim the lands without incident. By the 1880s, the U.S. had moved the last of the Horse Nations to reservations, an effort that combined both the extermination of the buffalo and the removal of the Horse Nations' horses.

The stockmen set out to rid the plains of wild horses using a few methods: bounty hunters to kill the horses, and mustangers to round them up and ship them off for sale. Homesteaders also used a simple product invented in 1873 that would have a major influence on the West: barbed wire.

Barbed wire closed off the open range. Fencing off the plains to control the movements of domestic and wild animals had enormous effects: it ended the romantic era of the nomadic cowboy by cordoning off the grasslands;[355] it pushed wild horses and other wildlife out of much of the plains and into rough terrain uninhabitable by most domestic livestock;[356] it caused overgrazing and denuding of the fragile grasslands by preventing cattle and other livestock from freely moving on to fresh grass; and it changed the nature of ranches and ranch work, as well as the types and quantities of horses needed to accomplish that work.[357]

Barbed wire also contributed to the cattle crash of 1886-1887 when blizzard conditions killed off 85 percent of the cattle in the West, as they could not travel beyond their fenced pastures to seek shelter or find food and water.[358]

As homesteaders unfurled barbed wire across the plains, railroads migrated closer to ranches, shortening cattle drives to market. Nomadic cowboys lost their lifestyle and turned into ranch workers. They no longer required a string of ten horses each to move cattle around a ranch or the short distances to new railway stations. They spent time fixing fences and maintaining windmills that pumped water for the cattle.[359] They also had to maintain and mow fenced hay fields, which required harness teams of horses. These horses had to be larger than the typical Spanish cow pony or Quarter Horse of that time, so ranchers brought in draft horses to breed with their mares to size up their stock.[360]

During the late 1800s, the Great Plains underwent a sweeping transformation. The arrival of the transatlantic railroad ushered in the heyday of long cattle drives across the open range. But between the takeover of the plains by absentee investors and the widespread use of barbed wire by settlers, the open range and the nomadic cowboy were swiftly fenced off and shut down by the end of the century.

28

MUSTANGERS

She was about fourteen years old but slight, maybe seventy pounds.[361] She raced her horse up alongside a wild horse and leapt onto its back. All she carried was a rope, which she quickly looped over its head, then twice around its muzzle. The mustang kept galloping, but within half an hour, the girl had gained control and ridden it back to her waiting family. They were mustangers, who had come up to Texas from Mexico on an annual trip to catch wild horses. They planned to trail them back to their ranch to sell.

For as long as modern horses have run wild in America, humans have tried to catch them. Early settlers were often in awe of the wild horses, who represented the freedom and ruggedness they themselves had come to find in this rough country. Some tried to catch mustangs just for the sheer rush of the chase. American-bred horses of the East were extremely expensive and difficult to come by in the West in the 1800s, so most early farmers and ranchers continued to stock their operations with mustangs to use as work animals.[362] Others rounded up mustangs to sell for profit. Still others shot them to collect a bounty.

Most mustangers were probably not as talented as the girl in the story above. They used a variety of methods to capture mustangs before mechanized vehicles were invented. Colt-catching involved chasing down a band of wild horses until a colt fell behind its mare.[363] Once the band disappeared, the colt would turn and follow the mustanger's horse.

Another method was to set up a snare by laying a looped rope near a watering hole, then cinching a horse's front legs together when it stepped into the loop.[364] Penning involved setting up a corral around a watering hole, then scaring up and herding a band into the corral. Another common method was just to chase a mustang down until it was too tired to continue fleeing.

Most early mustangers captured wild horses for their own uses, or for small-scale markets, and did not make an appreciable dent in the West's wild horse population. The market for wild horses ebbed and flowed throughout the 1800s and early 1900s. Just as the arrival of the transcontinental railroad transformed the cattle industry, it also greatly influenced the wild horse industry by opening up large eastern markets. Coupled with the fact that the investors who had taken over much of the cattle industry wanted to rid the pastureland of the Great Plains of wild

horses, mustanging became a profitable business in the second half of the 1800s.

Stockmen hired bounty hunters, often the same men they hired to exterminate wolves, to slaughter America's wild horses. (Wolves and mountain lions are natural predators of horses. It is ironic that westerners killed off the predators that could have helped to keep wild horse populations in check today.) In the late 1880s, a bounty hunter could earn $4 per pair of wolf ears.[365] But he could earn $25 for the scalp of a wild stallion.

During the peak of the cattle drives, approximately one million wild horses were trailed out of Texas alone and shipped east by railroad to sell.[366] Wild horses were wanted around the country for mines, farms, southern plantations, and city work. Overseas buyers bought mustangs for South Africa's Boer War. Mustangs furnished English, French, and Italian militaries during World War I.[367] Canning factories bought live horses to slaughter for dog food.

Once mechanized vehicles, such as trucks, airplanes, and helicopters, became available, the scale and cruelty of roundups increased dramatically.

When professional mustangers started to round up massive numbers of horses, they selected the highest-quality animals to sell and left behind the weaker ones.[368] Due to this practice, as well as the constant addition of escaped domestic horses into wild herds, the wild horse population began to degenerate from its strong Spanish origins.[369]

For two hundred years, and for millions of years before that, wild horses had roamed the American West, providing Native Americans and early settlers the stock they needed to survive and prosper. But by the close of the 1800s, professional mustangers, bounty hunters, and their clients had thrust the wild horse population, for the first time since the reintroduction of horses to North America, into decline.

29

FALL OF THE HORSE NATIONS: THE COMANCHE

In the decade leading up to the Civil War, the U.S. government had just reestablished how to use some of the Comanche's own battle techniques to fight against them. It seemed like the end was near for the powerful Horse Nation. But during the war, the government largely left the West to fend for itself. With no presence of military or even much in the way of local militia, the frontier descended into an even more lawless state than it already had been leading up to that time. Comanche brutality against the settlers and other native peoples went completely unchecked.[370] The Comanche raids on supply trains essentially shut down the Santa Fe Trail.[371] They also became cattle rustlers, in addition to horse rustlers.

The situation was dire. After the Civil War, the Comanche population was down to about four thousand, but they had basically halted westward immigration across the southern Great Plains.[372] Western homesteaders, stockmen, and miners called on the government to solve once and for all the "Indian problem," as they called it. They wanted the Great Plains free of Native Americans, the buffalo they followed, and the wild horses they helped to spread. The federal government redoubled its efforts, determined to wipe out the last remaining Horse Nations and reopen the frontier to migration.

The U.S. had mostly cleared indigenous peoples out of the desirable lands of the East by this point, but the Horse Nations of the West were another matter. Slowly, the government began to subdue the less aggressive of them, sending survivors to reservations. The country, however, had failed to control the most powerful of the Horse Nations; a region in the middle of the country still belonged to the U.S. in name only. With the prodding of the western settlers and investors, along with the newly freed up military resources, the government decided it was time to strike.

Perhaps ironically, many of the same generals, including William Tecumseh Sherman, who helped the Union win the Civil War and abolish the indignity of slavery, promptly turned their attention to eradicating native peoples of the West. These military minds realized the nomadic Native Americans of the plains were inextricably linked to their horses and the buffalo. Ridding the plains of these peoples, their horses, the wild horses they left in their wake, and the buffalo were not separate issues, but entirely interwoven; subduing the remaining Horse Nations and forcing them onto reservations would require both taking their horses and slaughtering the buffalo, their main food source.

Without buffalo to follow, the Horse Nations would have no motivation to remain nomadic, disrupting the country's division and redistribution of the plains. Without their horses, these nations would lose the ability to efficiently hunt any remaining food sources or fight effectively. And they would no longer be participating in the spread of wild horses.

Although Americans had been killing buffalo on the Great Plains to some degree for decades, wholesale slaughter occurred between 1868 and 1881, with the new transcontinental railroad ensuring transportation of hides to large markets.[373] Profits aside, Americans loved hunting buffalo just for the fun of it. Trains would even slow down when they encountered buffalo herds, so people could shoot them out the windows.[374] The entertainment was simply in the kill; they would leave the carcasses to rot. Eventually, the train routes were littered with unseemly buffalo carcasses, and trains would close the windows for discretion as they passed.

Hunters killed thirty-one million buffalo during that short span of time, nearly sending them to extinction.[375] By the end of the 1880s, only a few hundred buffalo remained on the plains.[376] While this slaughter was in part just one more example of pure greed in the face of newly accessible natural resources, it was also a political act.[377] The government did not order the slaying of the buffalo—

Americans took on this job quite willingly—but many government officials agreed it was an efficient method to subdue the Horse Nations and refused to stop the killings. The decimation of the Horse Nations' primary food source meant they had no hope of maintaining their nomadic existence across the Great Plains.

Although every Horse Nation, and likely every band within each nation, has their own story of being either cajoled or violently forced to abandon their nomadic ways for reservation life, most of those stories share another common element: the government's removal of all or most of their horses.

The U.S. government realized it would never fully control the southern plains as long as the Comanche still claimed them as their own. But the fall of the Comanche was neither peaceful nor swift, the final chapter of their nomadic existence ending in fits and starts.

In 1867, under government pressure, another several hundred Comanche signed the Medicine Lodge Treaty, agreeing to move onto the reservation in Oklahoma. Now nearly a thousand of the Comanche lived on the reservation, but their movements were fluid.[378] The treaty had failed to provide food for the reservation Comanche, leaving them starving in the winter. Many still left the reservation to hunt in the warm months, continuing to raid settlements as they went. They refused to learn to farm. Moving some of them to a reservation had done little to protect the frontier. In

1869, the majority of the remaining Comanche still lived off the reservation.

In desperation, the military brought in Ranald Slidell Mackenzie in 1871 to command the Fourth Cavalry on the frontier. Mackenzie was an impressive military mind out of West Point, who had gained attention for his fighting prowess during the Civil War.[379] Mackenzie's task: to bring down the Comanche. To succeed, he needed to employ the Comanche battle techniques Texas Ranger Hays had begun to crack open in the 1840s, which resurged in the late 1850s but had long since been forgotten along the frontier.

Instead of the age-old frontier strategy of playing defense or trying to chase down bands of Comanche to punish them for specific raids, Mackenzie went on the offensive. Rather than staying on the outskirts of Comancheria, he made it his mission to infiltrate Comanche territory, learn the landscape, and beat them at their own game. Again, it would mean imitating the nomadic ways of the greatest Horse Nation. Despite new rifles that shot twenty rounds per minute, Mackenzie understood his men would have a difficult time competing with horsed Comanche warriors in battle. The only way to beat the Comanche was to ambush them and steal their horses.

In one engagement, Mackenzie led six hundred military troops and twenty-five hired Tonkawa scouts deep into Blanco Canyon of Texas to search for Comanche.[380] But such a group was easy for the Comanche to spot. The warriors, led by the rising star, Quanah Parker, immediately raided the military camp and stole sixty-six horses, reminding Mackenzie whom he was up against.[381] After the skirmish, Mackenzie's men tried to pursue the fleeing Comanche, but they lacked the Comanche's strong, fresh horses and their knowledge of the landscape, and were unable to catch them after pausing to wait out a strong storm.

In a later encounter, Mackenzie traveled deep into Comancheria, located a Comanche village, and launched a surprise attack. He and his men killed fifty-two Comanche, took 124 prisoners (mostly women and children), and confiscated three thousand horses.[382] Taking this number of horses would seriously hobble the nomadic band. Mackenzie put the horses under the guard of the Tonkawa for the night. Predictably, the Comanche stampeded their own horses and managed to win most of them back.

Somehow Mackenzie had underestimated the Comanche's horsemanship abilities yet again. But this engagement taught him another lesson: beating the Comanche meant taking their horses for good. From then on, he shot the horses he confiscated from them.[383]

The Comanche caught a break for a few years from significant government assaults while the government sent Mackenzie to the Mexican border to stop Kickapoo and Apache raids.[384] But a series of small military encounters coupled with an extremely harsh winter of 1873-1874 pushed more Comanche to the reservation. By 1874, only about a thousand Comanche still roamed the plains, just three hundred of them warriors.[385] These warriors once again began to wreak havoc across the frontier in a final, desperate push to rid their territory of white settlers. That summer, they killed 190 settlers. The government wanted them gone.

Mackenzie came back with a vengeance that year to bring the last of the Comanche to the reservation. His confrontations had served as reminders of the inferiority of the U.S. Cavalry's horses and horsemanship skills compared to the Comanche's, but eventually they helped Mackenzie piece together a winning strategy. He had also begun to master life and movement through the plains in a way most other whites had not.[386]

He and his men eventually found a horse trail across the plains that led them into the heart of Comancheria.[387] They climbed to the top of a nine-hundred-foot cliff, looked down and saw five villages of Comanche, Kiowa, and Cheyenne. In a surprise attack that was part of the Red River War, they descended the cliff into Palo Duro Canyon and ravaged the villages. Although the number of casualties, four Comanche, was relatively small, the troops managed to take the majority of the possessions, food, buffalo robes, and supplies of most of the remaining non-reservation Comanche. Perhaps most significantly, they took 1,424 horses. Having learned his lesson, Mackenzie gave 340 of the confiscated horses to the Tonkawa, who had assisted in the battle, and ordered the rest shot.[388]

The loss of their horses and the entirety of their supplies crippled the Comanche. Most of them slowly made their way to the reservation that spring.[389] The last of them surrendered that summer.

By June 1875, the plains were clear of the Comanche, the most militarily powerful Native Americans in the history of the country.

JULIA SOPLOP

30

FALL OF THE HORSE NATIONS: THE NEZ PERCE

M eanwhile, on the Columbia Plateau of the Northwest, the Nez Perce posed a different problem. They were less warring than the Comanche and often traded their prized Appaloosas and other goods peacefully with pioneers journeying cross-country along the Oregon Trail.

But one enormous obstacle stood in the way of peace for the Nez Perce: gold. After the gold discovery in California in 1848, traffic along the Oregon Trail in the Northwest slowed as settlers peeled off to try their luck in California.[390] As that gold rush began to fizzle, however, prospectors started to spread across the Northwest in the early 1850s in search of their own gold fortunes.[391] Many of them ended up in Nez Perce territory.

The government had an interest in white settlers inhabiting these lands.[392] The borders of the Pacific Northwest were not yet clearly defined, and settlement

192

gave the U.S. a better shot at claiming the land rather than losing it to Britain; occupancy was much of the battle.

In 1855, in an effort to contain the Nez Perce and allow for more white settlement, the government convinced some members of the Nez Perce nation to sign the Treaty of Walla Walla, which granted the Nez Perce more than half their ancestral lands in Oregon, Washington, and Idaho. As with many other indigenous peoples, they did not have one central leader who had the power to make an agreement on behalf of the entire nation, but the government ignored that fact. The Nez Perce who signed the treaty were mostly people who had converted to Christianity and believed assimilating to European-American ways offered the best chance for their survival.[393] The "non-treaty" Nez Perce had no interest in entering into such an agreement with the U.S. government to give away any portion of their homeland or to assimilate.

This treaty, if the government had upheld it by keeping white settlers off the Nez Perce land, would actually have been a much better deal than the U.S. had given most Native Americans. But the government refused to stop encroachment from white settlers or quell the non-treaty Nez Perce.

The situation became untenable in 1860 when word got out that a prospector had found gold on the Nez Perce reservation.[394] Soon gold miners overran the land. The government did nothing. By 1862, around eighteen thousand whites were living in Nez Perce territory.[395] The mining towns that sprang up were rough, violent places. Clashes erupted between whites and Nez Perce, and continued long after the gold rush had simmered down but many settlers remained.

The situation was so volatile by 1863 that the government coerced a small group of Nez Perce to sign another treaty renegotiating the boundaries of the previous treaty and giving 90 percent of their reservation land to the government, leaving the Nez Perce with just a small tract of land in Idaho.[396] The non-treaty Nez Perce rejected this new agreement and stayed in their homeland.

Friction between white settlers and non-treaty Nez Perce was at an all-time high. In 1877, the government had finally had enough of the situation and demanded the entire Nez Perce nation move to the small reservation.[397] Eventually, the leaders of the non-treaty bands realized they had no choice. They decided to move to the reservation after one last traditional gathering in a camas prairie in Idaho.[398]

Some young warriors disagreed with their chiefs' decision, however.[399] They left the gathering and began to attack white settlers in revolt. Panic spread across the plateau. The rest of the non-treaty Nez Perce understood this violence cost them all the ability to make a peaceful transition to reservation life. Instead, they would need to flee for their lives.

One might think the U.S. Cavalry could quickly subdue a band that consisted of just 140 warriors, and 660 women, children, and elderly as they fled across some of the most rugged terrain in the country, beginning with the Lolo Trail across the Bitterroot Mountains.[400]

But one would be mistaken, because the Nez Perce had advantages over their pursuers; they set out on their flight with incredible horsemanship skills—the entire group, not just the warriors—and nearly three thousand sturdy, high-quality, grass-foraging Spanish horses.[401] Like the Comanche, the Nez Perce were pursued by the U.S. Cavalry, who were led by Union heroes of the Civil War. Again, the horses, horsemanship skills, and fighting abilities of the U.S. Cavalry paled in comparison to those of the Nez Perce.

For two and a half months and twelve hundred miles, in what became known as the Nez Perce War, the Nez Perce managed to hold off two thousand soldiers from five army units, engaging in numerous battles and skirmishes along the way.[402] Time and again, the Nez Perce evaded capture.

It was only forty miles from the Canadian border, which the Nez Perce sought to cross to join the Sioux, that the cavalry was finally able to surround and overpower the remaining 418 Nez Perce—many of whom were women, children, elderly, wounded, or ill[403]—and their eleven hundred remaining horses.[404]

Although the Nez Perce did not have a single chief who represented all the bands, one leader emerged during this time who would become forever linked to story of their flight: Chief Joseph. After some of the more vocal chiefs were killed in battle, Chief Joseph stepped up to lead the survivors and speak on their behalf.

His famous words have been misinterpreted throughout history as words of surrender: "Hear me, chiefs. I am tired. My heart is sick and sad. From where the sun now stands, I will fight no more forever."[405] He spoke these words to his fellow chiefs, not to the U.S. generals who pursued him, as so many accounts tell it, and he insisted he was not surrendering but rather agreeing to cease fighting so the Nez Perce could come to an agreement with the government to return peacefully to their ancestral lands.[406]

Chief Joseph never agreed to the removal of their horses when he stopped fighting, either.[407] But the military confiscated nearly all their remaining eleven hundred prized Appaloosas, slaughtering some, keeping some for the cavalry's use, distributing some to the Native American scouts they had hired to help track the Nez Perce (the military had bribed these neighboring peoples with promises of gifts of the Appaloosas for their assistance), and scattering and misplacing any record of the rest. The Nez Perce had lost their horses.[408]

Their Appaloosas had allowed them the chance of escape. In the end, however, the Nez Perce lost not only their land but also their Appaloosas. They had become a horse culture without horses. The government treated them as prisoners of war and shipped them around to numerous western reservations, all of which had poor living conditions, never allowing this particular group to return to their homeland, including the Nez Perce's small reservation in Idaho, or reuniting them with their horses.

While the story of the Nez Perce nation's claim to most of their ancestral lands ends here, the story of the Appaloosa does not. In 1937, sixty years after the flight of the Nez Perce, historian Frances Haines published an article about the history of the Appaloosa in *Western Horseman* magazine, imploring horse owners to prevent the breed's extinction.[409] The article caught the attention of many horsemen, including Claude Thompson, a breeder of Appaloosas.[410] The next year, Thompson and a few other horsemen established the Appaloosa Horse Club with the goal of reinvigorating the breed, which they estimated had only a few hundred remaining horses in the country.

By this time, the majority of the Appaloosas located to begin the registry had been crossed with Arabians and other breeds. Later, the registry would encourage the introduction of Thoroughbred and Quarter Horse blood into the Appaloosa breed, which means that although today's Appaloosa is still distinguishable by its distinctive coat patterns, it has changed quite a bit from the Spanish horses of the Nez Perce in size, conformation (body shape), and abilities.

The future of the Appaloosa had long been out of the hands of the Nez Perce. The nation watched as their horse enjoyed a resurgence but grew less and less recognizable as the animal they had originally bred. So in the early 1990s, they decided to try to reclaim their cultural heritage and provide economic opportunities for their people by developing a breeding program to recreate that animal.[411] They named it the Nez Perce Horse.

Some historians have suggested that the original Appaloosas may have had the blood of the Akhal-Teke, an ancient horse of Turkmenistan.[412] To slim down the modern Appaloosa and bring it closer to its original conformation and gentle disposition, the Nez Perce crossed Akhal-Teke stallions with Appaloosa mares, which also had Quarter Horse, Thoroughbred, and Arabian blood. Today's Nez Perce Horses are often buckskin or palomino. Some have the classic Appaloosa spot patterns.

The Nez Perce nation finally has a horse to call its own again.

31

FALL OF THE HORSE NATIONS: THE BATTLE OF LITTLE BIGHORN

T he downfalls of other Horse Nations followed similar patterns to those of the Comanche and Nez Perce: they maintained their freedom for years largely due to their talent as mounted warriors, but the U.S. eventually figured out how to subdue them using a combination of the Horse Nations' own fighting techniques, more powerful weapons, and, perhaps most significantly, the realization that domination required removing and slaughtering their horses.

In 1868, the U.S. had determined the Black Hills of South Dakota were worthless. They relinquished them, as well as part of Nebraska, by treaty to the region's native peoples, including several bands of Cheyenne, Arapahoe, and Lakota Sioux, some led by Sitting Bull and Crazy Horse.[413] But in 1874, gold was discovered in the Black Hills, and white fortune seekers began to trespass in droves on native land.

The U.S. Army sent General George Armstrong Custer to investigate the claims of gold. Custer confirmed the hills were packed with it and also packed, illegally, with white prospectors. Instead of chasing off the trespassers, however, the U.S. sent a commission to renegotiate the 1868 treaty and take back the land.[414] The Native Americans did not go for it. So the government suggested a lease for the mineral rights.[415] Again, the native peoples said no. The government decided to forcibly "purchase" the Black Hills instead.

By this point, some members of these nations had already moved onto reservations. Following the new Black Hills policy, the U.S. declared that all Native Americans in the territory who remained off reservations were required to report to the reservations immediately, by the end of January 1876, or they would be considered hostile and forced by the military to do so.[416] It was unrealistic or impossible even that the non-reservation bands, all spread out across the territory, could travel to the reservations in the middle of winter. They did not want to, either.

When they refused to return to the reservations by the deadline, the U.S. Cavalry came out in full force to compel them. The Great Sioux War of 1876 resulted, which involved one of the most famous confrontations between the U.S. government and the indigenous peoples of the West: the Battle of Little Bighorn, also known as Custer's Last Stand.

The battle began with a horse raid.[417] The U.S. soldiers sent in rustlers from the Arikara nation to stampede horses.

Chaos immediately ensued, and the battle erupted.

Between six thousand and ten thousand native peoples were camped together at the battle site. Despite the surprise attack, the seven hundred men of the Seventh Cavalry could not compete with the horsemanship skills of thousands of mounted warriors. The Sioux and their allies whipped the cavalry soundly, taking no prisoners. They killed Custer. They won the battle.

But the high-profile defeat only made the U.S. shift its strategy and barrel ahead. The government declared the Sioux and their allies had completely nullified their 1868 treaty by waging war against the U.S., despite the fact that the battle had taken place in self-defense.[418] The government took the tribal members on the reservation as prisoners of war, despite their lack of involvement in the conflict. The government also forced many of their chiefs to sign away all rights to the Black Hills and surrounding lands, threatening to take their horses and guns if they refused.[419]

In came Mackenzie, who had broken the Comanche and finally removed the last of them to reservation life. The chiefs had made good on signing away the Black Hills, but Mackenzie came to the reservation to arrest them and remove their horses and guns regardless.[420] The Sioux had become another horseless Horse Nation, living as prisoners of war.

Mackenzie then set out to hunt down any remaining Native Americans on the disputed lands off the reservation. With the help of the Pawnee, to whom Mackenzie had given some of the horses he had confiscated from the Sioux, he located the remaining war-weary Cheyenne, engaged in battle, burned their village, shot their horses, and sent the survivors fleeing.[421]

Meanwhile, Sitting Bull and his people eventually made it to safety in Canada. They returned to the U.S. five years later in surrender.

Beaten down and tired of running starved through the wilderness, Crazy Horse and his people surrendered to the reservation in April 1877. When rumor spread the following September that Crazy Horse was going to leave the reservation with his people and go back to their homeland, government officials promptly arrested and killed him.[422] The Sioux were then ordered to move to another reservation on the Missouri River.

By the end of the 1800s, the West had undergone a dramatic metamorphosis. The age of the Horse Nations, who had dominated the plains for centuries, had drawn to a close.

In 1893, historian Frederick Jackson Turner formulated the Frontier Thesis, which argued that the ever-shifting frontier had shaped American democracy, history, and culture: "The existence of an area of free land, its continuous recession, and the advance of American settlement westward, explain American development."[423] Now the frontier was gone. A new era was beginning.

The U.S. finally extended coast to coast, not just on paper but in actuality. The government had solved the "Indian problem." Americans had found ways to extract and sell the natural resources of the West. The railroad and telegraph had connected the country. And horses had played roles in just about every part of these sweeping changes. In the new era of the 1900s, however, their significance would plummet.

PART VII

THE EAST
MID- TO LATE 1800s

32

AFTERMATH OF THE CIVIL WAR IN THE SOUTH

F ollowing the Civil War, much of the South lay in ruins. Cities had been destroyed. Crops had been burned. A large percentage of the livestock had been killed. Hundreds of thousands of southern men had died in battle or from disease. Horses still powered the country, and the South felt the loss of their horse population heavily. The southern agricultural economy, which had depended on slave labor, had been decimated. Formerly enslaved blacks faced an uncertain future in a region that resented them.

In short, the South was a mess. The years that followed, known as the Reconstruction Era, focused not only on rebuilding the region but also on defining whether blacks would have civil rights in the U.S. The Fourteenth Amendment, ratified in 1868, did grant citizenship and equal protection to all persons born or naturalized in the U.S., including the formerly enslaved. The place African-Americans would be allowed to hold in society, however, remained a bitter controversy.

Prior to the war, plantation owners had grown wealthy from their rice plantations in South Carolina and Georgia's Lowcountry and on the states' Sea Islands. The area was prone to hurricanes and full of malarial mosquitoes, so plantation owners left the management to others. What followed was the development of the Gullah community, a distinct culture with a creole language formed by a large number of enslaved Africans, many from Muslim countries, who lived relatively isolated from whites.[424]

The Sea Islands were the first to free enslaved blacks during the Civil War, and these islands became a refuge for others escaping slavery.[425] After the war, the government began to experiment on the Sea Islands with confiscating land from white plantation owners and granting forty acres per family and, supposedly, a mule to the formerly enslaved. When President Andrew Johnson took office, he quickly canceled the program before it expanded and gave the land back to the white plantation owners.[426] Although this failed program is referred to as "Forty Acres and a Mule," oral history tells us the government actually gave the Gullahs Marsh Tackies, the Spanish horses that ran wild on the Sea Islands, to work their farms.[427]

The Gullah people remained in the Lowcountry and on the Sea Islands, despite the failure of the government to make reparations for slavery. Some were likely sharecroppers, meaning they continued to work the land of the plantation owners and were required to hand over a large share of the profits to the owners. Some white land owners eventually abandoned their plantations after hurricane damage, and Gullahs claimed ownership of some of these parcels, continuing to farm the land. Marsh Tackies powered their agricultural work and transportation.[428] In addition to the Marsh Tackies they may have received from the government, the Gullah people rounded up more of these horses, which were still abundant in wild bands on the islands, when needed.[429]

Meanwhile, the war had devastated Virginia's horse population. Southerners had ridden their Quarter Horses and Thoroughbreds to war and lost many of them. Northerners had taken much of the rest of the state's horses for their own uses during the war. Virginia's breeding industry, which had started to lose its sway to Kentucky after the Revolutionary War, struggled. Kentucky, on the other hand, began to make moves to raise its horse-breeding status to an entirely new level.

Horses temporarily elevated the status of some African-Americans in the South following the war. Like black cowboys of the West, black jockeys and horse trainers of the South held a higher position in society than other blacks, at least for several decades following the Civil War.[430] They were the skilled workers who held the keys to producing winning race horses. For several decades, black men continued to dominate jockeying and training and earn respect for it. Isaac Murphy, a black man born in Kentucky during the Civil War, became the highest-paid jockey in the

country in the 1880s, raking in an estimated $20,000 per year,[431] the equivalent of about $534,000 today. But in the early 1900s, segregationist laws essentially banned black jockeys from racing.[432] White men leading the racing industry then proceeded to erase from its history the legacy of African-Americans who contributed to the sport.[433]

Despite substantial losses to its horse population during the Civil War, the South still relied heavily on the horse during the Reconstruction Era. Marsh Tackies helped the Gullah people to farm their lands. Horsemanship skills elevated the social status of black jockeys and trainers. The devastation of Virginia's horse stock during the war solidified its loss of status as a Thoroughbred breeding center and paved the way for Kentucky to earn the honor. While the South floundered through Reconstruction, the North flourished as industrialization blazed ahead, powered by the horse.

33

HORSES IN THE CITY

Nowadays, horses conjure up images of country living. But throughout the 1800s, cities were packed with them. Before the use of electricity and the invention of the combustion engine, which led to the development of the automobile, horses powered just about all aspects of city life from transporting people, to hauling goods, to fueling early manufacturing. Streets teemed with horses; the din of hooves and cart wheels and bells added to the chaos of urban life; the stench of manure permeated everything.[434]

In the decades following the Civil War, the country charged into its Second Industrial Revolution, meaning people were rapidly learning to build machinery that could produce large quantities of goods at a time and handle many jobs previously performed by hand. Industrialization caused cities to swell, especially in the North, and the horse population exploded to meet urban demand.

In 1860, there were approximately seven million horses

spread across the entire country.[435] The horse population took a large hit during the Civil War but rebounded relatively quickly due to the demands of urbanization. By 1900, the country had around twenty-five million horses, mostly in the East.

City horses accounted for about 12 percent of the country's equine population, and you could find just about any type of horse in cities, fueling industrialization.[436] Massive Shires and Clydesdales hauled freight wagons to and from railroads on the outskirts of town.[437] Handsome trotting horses pulled carriages. Saddle horses wove riders through the disorder of urban traffic. Heavy horses pulled fire engines and walked on treadmills to power equipment at lumberyards and brickyards.[438]

The highest proportion of urban horses pulled streetcars.[439] The earliest version of the streetcar, introduced in New York City in 1829, was the omnibus, a tall, unwieldy vehicle packed with people. Wild omnibus drivers added to the pandemonium of the street.

In the 1850s, the streetcar, also known as the horse railway because horses pulled it along a track, replaced the omnibus. These streetcars increased the geographic spread of cities, which led to the concept of commuting to work. For the first time, city dwellers could live well out of the fray but still access their urban jobs.

When the transcontinental railroad opened up shipping from the West, mustangers rounded up wild horses for profit, piled them in cattle cars, and sent them to work in

New York City.[440] By the end of the 1800s, there were about five times as many formerly wild mustangs pulling streetcars in New York City as there were bred horses from the East. The mustangs required relatively little food and care compared to larger, more sensitive horses. However, runaway mustangs were not uncommon, as they chafed against city living.

Just as horse-centered economies materialized around railroad stations, they also emerged in cities to care for and profit from urban horse populations. Livery stables rented out horses and horse-drawn vehicles and offered stabling for visiting horses.[441] Every neighborhood had its own stables. Farriers, stable hands, carriage makers, breeders, harness makers, and workers in numerous horse-oriented professions found work in cities.

The Industrial Revolution of the late 1800s depended on horse power, and the exploding horse population needed to eat. This requirement created a domino effect on the growth of other horse-related industries outside cities, such as growing hay and oats for horse feed.[442] Farmers could not keep up with the demand without advances in agricultural machinery.

You can only plant as much as you can harvest before the product goes bad. The invention of mechanized plows, reapers, and threshers for large-scale planting and harvesting helped farmers keep up with exploding demand.[443] Teams of horses, lots of them, drove this new machinery.

The heavy new machinery put an end to the use of oxen for farm work; although oxen were strong, they were too slow to pull it.[444] The need for more work horses continued to increase the demand for draft horse breeders.[445]

In 1872, when an epidemic of horse flu known as the Great Epizootic swept through the U.S. and Canada, infecting just about every major city within a year, urban dwellers became aware of how vulnerable they were to rely so heavily on horses.[446] Cities turned chaotic. The virus infected approximately seven-eighths of Boston's fifty thousand horses in just forty-eight hours.[447]

Although the flu killed only about 1-2 percent of afflicted horses, it took sickened horses out of commission for at least a few weeks, meaning nearly every industry was left without work horses.[448] For example, a major blaze in Boston wreaked havoc when the fire brigade's horses were too sick to haul their fire engines.[449] The epidemic demonstrated just how dependent urban life and industries were on the horse, and how devastating a more lethal virus could be on the economy and daily living.

This epidemic, as well as others that had spread through horse populations during the Civil War, pointed to a need to look beyond horses for alternative forms of energy. Those forms of energy have obviously proven to have graver health, environmental, and economic consequences than horses, though.

The epidemics also demonstrated a need for a profession that provided medical care to horses, which resulted in the establishment of the first formal, accredited veterinary schools in the country.[450] Their main objective was to train professionals to care for urban horses. Their primary mission was to protect the economy.

While horses served important roles in urban economies, they were also nuisances. Their stables took up prime real estate. Their manure covered roadways, contributing to public health concerns and creating a logistical nightmare to clean up. In 1900, New York City's 131,000 horses produced thirteen hundred to thirty-three hundred tons of manure per *day*.[451] Horse carcasses littered the streets. Their unpredictability on crowded roads also resulted in injuries to humans and horses.

While there were undoubtedly people who treated their horses humanely during this era, many treated them more like machines than animals. By the end of the century, large freight corporations had bought up many smaller businesses, which meant that many of the people handling horses on the street were not their owners, but rather employees of the companies that owned them.[452] They were not animal lovers invested in the health and safety of the horses they drove; they required horses to get their work done and were often abusive to them. Anna Sewell's 1877 novel, *Black Beauty*, which lobbied for more humane treatment of horses, contributed to shifting attitudes about how horses deserved to be treated at the close of the century.[453]

In 1890, horses pulled up to 90 percent of streetcars.[454] But while the 1890s were a peak of urban horse use, they also signaled the beginning of the end of horse-powered society. Along came cable cars—streetcars pulled by cables and powered by steam engines in central power stations, which were good for hauling streetcars up steep hills but broke down easily. Then came the electrification of the streetcar. These innovations made it clear that society was inching toward finding energy sources beyond the horse. The fate of the urban horse, which had powered the 1800s, would be just about sealed shortly after the turn of the century. By 1902, electricity fueled about 97 percent of streetcars.

34

HORSE RACING AND THE REBRANDING OF KENTUCKY

D uring the late 1700s and through the 1800s, Americans continually carved up and reshaped the country's landscape to make it more efficient for horse power. At least one state went so far as to remake not only its landscape but its entire image to win back the horse breeding industry.[455]

Dominance in horse breeding and racing had shifted between the North and South ever since colonists in the East started to amass enough wealth to own and breed horses specifically for racing.[456] However, the Panic of 1837, a national financial crisis that spurred a major recession, along with anti-racing and anti-gambling laws in New York, all but shuttered breeding and racing in the Northeast leading up to the Civil War. At that time, Kentucky held firm control of Thoroughbred breeding.

When the war began, Kentucky was a slaveholding state, but it was not considered part of the South.[457] Rather, it was considered western and had quite a reputation for lawlessness.[458] About 100,000 Kentucky men fought for the North, and between 20,000 and 45,000 fought for the South.[459] But most refused to fight altogether.[460] Regardless, the Civil War ravaged the state. Both armies, as well as outlaws looking to make a buck selling to the armies, raided its high-end horses, leaving Kentucky's breeding and racing industry in a shambles by the end of war.

Meanwhile, New Yorkers were rapidly accumulating wealth, thanks to industrialization. Racing had always been a way to demonstrate social status, and now that New Yorkers had cash to burn, they developed a desire to revive the racing scene in the North.[461] The Saratoga Race Course opened in New York in 1863, right in the middle of the Civil War.

These affluent New Yorkers initially sourced their Thoroughbreds from Kentucky.[462] But as the war dragged on and began to hamper breeding operations in Kentucky, they realized they could take the sport into their own hands.

The New Yorkers built their own extravagant Thoroughbred breeding farms to supply the North's reinvigorated racing scene.[463] Kentucky risked losing its horse-centered economy altogether.

And then after the war, something incredible happened: Kentucky managed to rebrand itself, and in doing so, regained its dominant position as the center of Thoroughbred breeding.

Here is how it went: before the war, Kentucky was not

part of the extravagant South that Margaret Mitchell portrayed in her famous novel, *Gone with the Wind*. The state had established a prominent breeding industry, but its countryside was not concentrated with massive plantation mansions and the enormous wealth to match found in the literature of the pre-war South, though there was certainly some of each.[464] It was more like the Wild West—rough-and-tumble, dangerous.

After the war, the few remaining breeders in Kentucky realized the state could not rebuild itself to compete with the new industrialist money in the Northeast.[465] The only way to reestablish dominance was to convince wealthy New Yorkers to relocate their breeding operations to Kentucky.[466]

At the same time, plantation tradition literature was resurging in response to Harriet Beecher Stowe's anti-slavery novel, *Uncle Tom's Cabin*, and to the Civil War. This genre of literature, often first credited to Thomas Nelson Page's *In Ole Virginia*, romanticized the bygone Old South and the institution of slavery itself, framing plantations as loyal families with black and white members, and suggesting enslaved people were happy to be so and better off that way.[467] Ironically, northerners began to consume this literature after the war and buy into the glamorized plantation lifestyle—the same wealthy agricultural economy and culture built on slavery that the North had just abolished.[468]

Novels by Kentucky authors, including Annie Fellows

Johnston and James Lane Allen, cast the state as a part of this Old South rather than the West, which began to shift the country's image of it.[469] A few brilliant individuals trying to rebuild Kentucky's horse industry noticed this attitude and jumped at the chance to capitalize on it to remake Kentucky's image from a backwoods, lawless, western state not stable enough for investment into a southern paradise to attract northern wealth.

Kentucky's promoters began to tout the state's new, mythical Old South image, along with its rich bluegrass, through various northern business interests and periodicals, such as *Turf, Field and Farm* and the *Live Stock Record*.[470]

In 1885, Kentuckians caught a break when one of the most prominent names in New York racing, August Belmont, moved his Thoroughbred breeding operations to Kentucky. Little by little, more northeastern horse breeders set up shop in Kentucky, buying up smaller farms and combining them to build extravagant breeding operations, which attracted even more wealth and investment, transforming Kentucky into the very image of the Old South its horsemen had been trying conjure up and sell.[471]

Kentucky's rebranding from a rough-and-tumble western state to the high-class Old South was a long shot, but it worked. Kentucky's Thoroughbred industry began to boom once again, solidifying it as the breeding and racing powerhouse it remains to this day. Today, the state's equine industry has a $3 billion impact on the economy.[472] It is also home to one the most famous race tracks in the world, Churchill Downs, and hosts one of the most famous races, the Kentucky Derby. The horse defined Kentucky's early economy, and the state reshaped its image to reclaim and maintain that economy.

PART VIII

A UNIFIED NATION
1900s-2000s

35

THE AUTOMOBILE

Through the 1800s, most improvements in transportation, such as artificial roads, canals, railroads, and streetcars required horse power to work efficiently, thereby increasing the demand for horses. Then along came what was likely the most revolutionary transportation innovation since the domestication of the horse five thousand years ago: the mass-produced automobile.

The "horse-less carriage," as the automobile was known, appeared at the end of the 1800s. But in 1908, Henry Ford released the assembly-line-produced Model T. Suddenly everyone had to have one. The rapid acceptance and spread of this new technology happened for several reasons, all related to the horse. For starters, the automobile did not require the construction of new infrastructure; unlike the train, barge, or streetcar, it could run on existing roads that had already been developed—for the horse.[473] Social pressures from advocates of humane treatment for horses and the publication of *Black Beauty* had also convinced many people to rethink how much work was acceptable to require of an animal.[474] In addition, a more recent understanding of the vulnerability of businesses to equine epidemics contributed to an eagerness to adopt new transportation, as did concerns about public health and increasing real estate values.[475]

The horse population for non-farm use in the U.S. peaked in 1910.[476] Between 1910 and 1920, however, horse populations in cities declined by half due to the uptake of the automobile.[477] In New York, the number of cars surpassed the number of horses on the road by 1912.[478] Horses still did urban hauling through the 1920s and beyond though, and the city of Philadelphia even continued to use some work horses into the 1950s.[479]

The speed with which the automobile replaced the horse happened more slowly outside cities and across the West. Horses were still considered the best energy sources for most agricultural work until the 1930s, when tractors and other motorized machinery became more widely available and started to win out over horses on farms.[480]

The decreased demand for horses during the first several decades of the 1900s had a widespread effect on all horse-related industries: breeders shuttered their operations, veterinary schools boarded up their doors, farmers growing grain and hay went broke.[481] In fact, *The Farm Horse*, a 1933 report by the U.S. Department of Commerce and Bureau of the Census, called the damage to the agricultural sector from the loss of demand for horse feed "one of the main contributing factors" to the Great Depression.[482]

The combustion engine that powered the automobile was not the only innovation that reduced the need for the horse. At the same time, electricity was spreading throughout the U.S. The concept that you no longer needed to own all your energy sources, and feed them and house them and care for them, but could now use energy at a different location from where it was produced, swept across the country.[483] By 1902, almost all streetcars had been electrified.[484] Innovations in electrified machinery of all sorts followed, causing horse populations to decline further.

Due to the adoption of the automobile and electricity during the first half of the 1900s, the horse transitioned from being the widespread power source of the industrial and agricultural revolutions back to, with some exceptions, a status symbol of the wealthy. This transition had huge ramifications for horse-related industries, as well as for the future of America's wild horses.

36

THE DECLINE OF THE DOMESTIC AND WILD HORSE

A s demand for the horse as an energy source began to drop with the spread of the automobile and electricity, two more circumstances further forced the U.S. horse population into freefall: World War I (WWI) and the invention of mass-produced, canned dog food.

Almost as soon as WWI broke out in Europe, it became apparent that horses were no match for modern weaponry, such as machine guns and mustard gas. They were not used much for cavalry charges, but millions of horses still hauled weapons and supplies during the WWI, much as they had during the Civil War. They also served as a food source if they were killed in service.

The U.S. shipped nearly two hundred thousand horses overseas with its troops during WWI and exported an additional million horses to Europe throughout the war years.[485] Some of these animals were domestic horses displaced from service by the automobile, but many of them were wild. Arranged by middlemen, mustangers would round up the wild horses from the West, herd them to railroads, and send them to waiting ships, turning a profit at a time when the horse trade had suddenly become nearly worthless otherwise.

The U.S. Army Remount Service, which sought to supply and resupply troops with horses, even turned some of their domestic cavalry stallions loose in the West to breed with wild horses to create hardier offspring than their domestic horses could provide.[486] They would then round up the wild offspring for military use and ship them to Europe. (Some offspring likely remained in the wild.)

These exports chipped away at both wild and domestic horse populations. Only two hundred horses returned to the U.S. from Europe after the war.[487]

The shrinking demand for horses due to the development of mechanized vehicles meant wild horses came to be seen as nothing but a nuisance, no longer offering any benefit to humans while also eating up prime grass that cattle or sheep could have been grazing on.

Humans had found ways to profit from mustangs to some degree for as long as horses had been running wild out West. To continue profiting as demand for work horses fell, they simply had to pivot. Both the railroad and mechanization made such a shift possible. Americans mostly found horse meat distasteful for human consumption, but it turned out that for a while, they did not view horse products as offensive for other uses.

In the 1920s, P.M. Chappel developed the first canned dog food.[488] His prime ingredient? Horse. Chappel built a cannery near a railroad line in Illinois and hired mustangers to round up wild horses out West and load them onto trains headed for his slaughterhouse.[489] Mechanization made it possible to mass-produce cans of dog food, which meant Chappel needed tons of horses. Although chicken feed, fertilizer, and soap companies also got into the wild horse game, it was Chappel and his dog food company that removed the most wild horses from the American West, producing six million cans of dog food per year at the height of production.[490]

The combination of the war, lowered domestic demand for work horses, the broad reach of the railroad, and mechanized factories that could mass-produce goods had a significant result: the wild horse population in North America, which had been in decline for decades, began to plunge.

By the late 1920s, the West's most accessible wild horse herds had been shipped off to slaughterhouses.[491] And then, just as the profitability of mustanging was being questioned, along came the airplane, and later the helicopter, capable of scaring up bands from the most isolated, desolate locations.[492] As early as 1938, pilots would locate bands and herd them from the air into specially constructed corrals, a substantially more efficient method than just using horses or trucks to force the animals out of the bush, but a more traumatic one for the horses, too. With few to no legal protections, the mustang population continued to drop precipitously. During their peak in the early 1800s, several million wild horses had roamed the West; by the 1950s, only an estimated twenty thousand remained.

37

WORLD WAR II AND THE LIPIZZANER RESCUE

Y ou would think after seeing how poorly horses fared against the weaponry of WWI, the U.S. military would have given up its cavalry altogether in the wake of the war. But that was not the case. The U.S. Army Remount Service was still actively acquiring and breeding horses as World War II (WWII) approached.[493] Die-hard horsemen heading up the cavalry were sure the horse still had much to contribute to the next war effort.

In 1941, the U.S. held its largest war games ever to test out its military technology, such as tanks and other motorized vehicles, and to determine whether the mounted cavalry could still play a significant role within the mechanized military as it prepared for war.[494] During these war games, known as the Great Louisiana Maneuvers, the U.S. assembled 470,000 men, 50,000 wheeled vehicles, and 32,000 horses.

The troops were divided into two teams: one with

motorized vehicles only, and one with horses supported by motorized vehicles. Eventually, the team with motorized vehicles overtook and surrounded the team with horses, demonstrating the efficiency of a horseless, fully motorized military. As a result, the military deactivated the cavalry and did not send horses overseas with the troops to fight WWII, with the exception of one unit.

The Twenty-Sixth Cavalry was dispatched to the Philippines, where it participated in the last cavalry charge in U.S. history—that is, until 2001, when several Green Berets joined with the Afghanistan Northern Alliance and entered into mounted combat.[495] The Twenty-Sixth Cavalry held off Japanese tanks as the American-Filipino unit retreated to Bataan, but they were devastated in the process.[496] Those who survived the engagement were later forced to eat their mounts to avoid starvation.

Some of the thousands of horses used in the Great Louisiana Maneuvers that remained state-side during the war were likely set loose or escaped, joining local wild bands.[497] Today, descendants still live at Fort Polk and Pearson Ridge Military Training Area, where they are considered "trespass" animals, and the military has been rounding them up and removing them. These animals are not subject to protection under more recent laws.

On the home front during WWII, Marsh Tackies, which performed well on soft, mucky terrain, helped to patrol the beaches of South Carolina, their riders watching the water for German U-boats.[498]

Immediately following the war, the U.S., as the dominating force of the United Nations Relief and Rehabilitation Administration, repurposed war ships to send draft horses and other livestock to Europe to help rebuilding efforts.[499] Much of the continent and its crops had been destroyed, and many of its horses killed. Using draft horses was a more affordable and realistic way for struggling farmers to replant their crops, rather than trying to acquire expensive, newly developed tractors. "Seagoing cowboys," often provided by the Church of the Brethren, traveled on the ships to tend to the animals.[500] This program grew into today's Heifer International.

Perhaps the most fascinating story of U.S. involvement with horses during WWII, though, is the most unexpected one. During the war, the Nazis set out to create not only what they called a master Aryan race, but also a master equine breed to serve as warhorses.[501] Germany used 2.7 million horses during the war, mostly for hauling supplies, and the leader of the selective breeding operation believed the perfect horse would be able to outperform any mechanized military vehicle. To breed their ideal horse, the Nazis located and confiscated what they viewed as the best—and whitest—horses from across Europe, which were mostly Lipizzaners, the world-famous dancing horses of the Spanish Riding School in Vienna, as well as some purebred Arabians.

While the Nazis were packing humans into cattle cars and shipping them to concentration camps, they treated these prized animals with special care, giving them personal attention and transporting them in luxurious, padded train cars.[502]

By late 1942, the Nazis had collected almost every Lipizzaner in the world, with the exception of the riding school's actively performing stallions, and were holding them in one location: a breeding farm in Hostau, Czechoslovakia.[503] It was a dangerous strategy in the middle of war; one air strike could take out the entire breed.

As the war progressed, and the Nazis' position grew more precarious, they began to worry about the fate of the horses. By the spring of 1945, the Americans were closing in on German territory from the west, and the Russians from the east. Unlike the Nazis, the Russians were brutal when it came to horses.[504] Their armies were also starving. If they got to Hostau before the Americans, there was a good chance they would shoot and eat the Lipizzaners, Austria's national treasures and now the prized possessions of the Nazis.

With the Russians within a few miles of the farm housing the Lipizzaners, the horses' principle caretakers, including Germans who wore the Nazi uniform, knew they had but one choice to save the animals' lives: they had to ask the Americans for help.

What transpired was almost unthinkable. When a German soldier arrived at the nearest American outpost to ask the Americans to evacuate the horses to safety, it happened to be the camp of Colonel Hank Reed upon which he stumbled.[505] Reed was a former cavalry officer who had devoted his entire career to horses until the moment they were stricken from the country's war strategy following the Great Louisiana Maneuvers. It also happened that Reed's men were actively looking for this particular German officer and his entourage of spies, who were holed up nearby.

Reed was perhaps one of the few officers in the U.S. military who could fully appreciate the significance of these particular horses, and who would be willing to put lives on the line to rescue them from the approaching Russians. He was also one of the only officers close enough to have a chance of saving them. Reed managed to acquire the secret, reluctant permission of his friend, General George Patton, for the mission, though Patton said he would not take responsibility if it ended poorly.[506] In exchange for the surrender of the German officer and his fellow spies, Reed agreed to try to rescue the horses.

The mission, "Operation Cowboy," was full of danger. The horses' German veterinarian traveled behind American lines to help arrange the mission, which involved the Americans staging a fight with the Germans and their prisoners of war, who helped run the farm, to make it look like they had no choice but to surrender.[507]

There were Germans unrelated to the mission sprinkled throughout the woods surrounding the farm, trying to hold off the American forces as long as they could.[508] The Russians were advancing, and the Americans had to move beyond the border they had agreed upon with the Russians to access the horses. Gunfire ensued on several occasions, and an American soldier was killed during the mission.[509]

Once the Americans had secured the horses and their prisoners of war, removing the animals from Czechoslovakia became a logistical nightmare.[510] They had to find vehicles to transport the weakest horses, and enough soldiers and prisoners of war with horse experience to herd the rest to safety. They also had to fight with Czech border guards to allow them to take the horses out of the country and into Germany on their route to Austria. Then they had to find land and stables to board the horses, along with feed for them in a country that had been devastated by years of war and could barely feed its people, let alone its animals.

They were able to return 218 Lipizzaner breeding mares to Austria to the Spanish Riding School.[511]

Separately, the director of the school, Alois Podhajsky, had miraculously managed to whisk the seventy Lipizzaner stallions who lived and performed in Vienna out of the city as it came under fire, seeking refuge in rural St. Martin, Austria.[512] In another twist of fate, Podhajsky was able to organize a performance for Patton in St. Martin. Patton, a foxhunter and polo player himself, was so impressed that he officially granted the Spanish Riding School protection by the U.S. Army.[513] Now the breeding mares of Hostau and the performing stallions of Vienna all had U.S. protection. The Spanish Riding School would survive the war.

The Nazis had also acquired about forty Lipizzaners and a few Arabians from other places in Europe, including Italy and Yugoslavia.[514] In a complicated twist, these horses ended up as spoils of war and were shipped back to the U.S.

When people later asked why, in the face of so much human horror, he decided to expend resources to save the horses, Reed said, "We were tired of death and destruction. We wanted to do something beautiful."[515] Austria's Spanish Riding School still operates today in Vienna.

The U.S. Army largely left its horses at home during WWII, finding them inefficient compared to mechanized vehicles. But a respected cavalry officer and his men recognized the cultural value of Austria's treasured Lipizzaners. They risked their lives and reputations to collaborate with an enemy and save the animals.

38

WILD HORSE ANNIE AND MUSTANG PROTECTIONS

For the time and the place, Velma Johnston was an unlikely character to launch a movement to save America's wild horses. For starters, she was a woman in the man's world of ranching and politics. And though she had grown up around horses, she had spent most of her career in Nevada behind a typewriter, not in a saddle.

One day in 1950, however, Johnston noticed something strange on her drive to work: a truck speeding along the road, leaking blood and packed with injured mustangs.[516] On a hunch, she followed the truck. When it came to a stop, she asked the driver where he was taking the horses. He said the mustangs had been rounded up from the Virginia Mountains and were headed to a slaughterhouse.

Horrified, Johnston decided that day it would be her mission to protect the country's last remaining mustangs from slaughter. She spent the next two and a half decades fighting for the rights of America's wild horses, eventually winning her battle, at least to a degree. Her success would have major consequences for land and horse management in the West that continue today.

A wild stallion chases off bachelor stallions harassing his band.

At the time Johnston happened upon the truck of bloodied horses, there were only an estimated twenty thousand mustangs left in the West.[517] They had mostly been fenced out of prime grazing lands with the advent of barbed wire and pressed into the most desolate areas, where they could survive on forage too nutrient poor for most other animals, due to their adaptation of the large cecum.

By then, some states and counties in the West had laws regulating how residents could go about rounding up or killing mustangs. Others had none. Regardless, it was basically open season on mustang hunting. In fact, the Bureau of Land Management (BLM), the federal government agency tasked with handling livestock grazing on public lands, even hired mustangers to round up and send wild horses to slaughter.[518]

Johnston realized it was now or never to save the horses that had once been a symbol of liberty and whose history was interwoven into the country's own. She became a wild horse advocate, showing up at local government meetings to fight against roundups. She also launched a grassroots movement to push for state legislation by sending letters to anyone she could think of who might be interested in protecting wild horses: newspapers, schools, 4-H clubs, humane societies.[519] Over time, her efforts helped to sway public opinion.

Her first success came in the way of a county ban on aerial roundups, which often result in horse injuries and deaths. Then in 1959, Congress passed the first wild horse protection act in Washington, banning aerial roundups on federal land. The bill, however, had its limitations. It did not prevent roundups by other means, or sale for slaughter.[520] In addition, the government did little to enforce the law, so aerial roundups continued largely unimpeded.[521]

Johnston's movement continued to grow over the years, and her political victories brought even more attention to the cause. Along the way, someone gave her the nickname Wild Horse Annie as an insult. The name stuck, and Wild Horse Annie became the unlikely face of the movement. The public demanded to preserve America's wild horses.

Around the country, activists began to fight for stronger protections for their local mustang bands. The dry, desert-like Pryor Mountains, which straddle the border of Montana and Wyoming, were home to a population of just over two hundred wild horses. The horses predated the arrival of white explorers and settlers to the area and are thought to be descendants of the Spanish horses of the Crow nation, who possessed the most horses of any northern Horse Nation.[522]

The nearby community in Lovell, Wyoming, had long respected this herd of horses, considering them a piece of the region's heritage.[523] So the community was distraught when the BLM announced in 1964 that it would round up and remove the herd, or at least most of it, citing erosion of the range. The residents became vocal advocates for their local wild horses and managed to hold the roundup off for years while they made their legal case against it.

The Montana Fish and Game Commission, the advocates learned, contended the horses had overgrazed the landscape, thereby reducing the population of mule deer, which people liked to hunt.[524] In addition, the Commission wanted to rid the land of wild horses, so it could build up a population of bighorn sheep for hunting. The Commission was also working to purge the entire state of wild horses. Unbranded horses were technically the property of the states' stockmen's associations in Montana and Wyoming at that time. The advocates believed these hunting and livestock interests were pressuring the BLM to unduly blame the wild horses for the state of the range as an excuse to remove them.

Wild horses of the Pryor Mountains of Montana.

A national uproar followed. In 1968, after legal battles, the Department of the Interior, under which the BLM falls, backed off the horse removal plan and surprised the public by establishing the Pryor Mountain Wild Horse Range, the first of its kind on public land in the U.S., dedicated primarily to providing a safe territory for wild horses.[525] It now encompasses nearly forty thousand acres.

Despite the victory for wild horse advocates, the BLM decided to appoint an advisory committee that encompassed representatives from both sides of the conflict to evaluate the range; determine whether wild horses had degraded the landscape; and recommend a long-term management strategy of the range and the wild horse herd, as well as the rest of the range's wildlife, including mule deer.[526] A range management specialist analyzed the land in question and determined that domestic sheep had caused the erosion fifty to a hundred years prior, exonerating the horses. The horses were not in danger of starvation, either. In fact, they were making a living where other animals had failed to.

Wild horses graze and socialize in the Dryhead area of Montana's Bighorn Canyon, part of the Pryor Mountain Wild Horse Range.

In 1969, the advisory committee made the unanimous recommendation that the land be reserved for wild horses above all else, since they were the animals best adapted to live in the Pryor Mountains.[527] More recent genetic analysis has confirmed that the Pryor Mountains herd is one of the purest Spanish horse herds in the country, lending further evidence to the cultural heritage of these horses.[528] (In 1998, the Pryor Mountain Wild Mustang Center was established in Lovell. It serves as home to the area's advocacy group, which remains committed to protecting their local wild horses and collaborating with the BLM to do so.)

The national fight to preserve wild horses was well underway. In 1966, Marguerite Henry, the author of *Misty of Chincoteague*, published a fictionalized account of Velma Johnston's life and work called *Mustang: Wild Spirit of the West*. The book, along with Johnston's own efforts, spurred children to flood their congressional representatives with letters, begging them to support measures to save the wild horses. Their pleas worked.

In 1971, Congress passed a bill, written and lobbied by Johnston, that changed the way the federal government would manage wild horses on federal lands to this day. Some Congressmen said it was the children's letters they received that convinced them to support the bill. It begins:

Congress finds and declares that wild free-roaming horses and burros are living symbols of the historic and pioneer spirit of the West; that they contribute to the diversity of life forms within the Nation and enrich the lives of the American people; and that these horses and burros are fast disappearing from the American scene. It is the policy of Congress that wild free-roaming horses and burros shall be protected from capture, branding, harassment, or death; and to accomplish this they are to be considered in the area where presently found, as an integral part of the natural system of the public lands.[529]

The act refers to these horses as "wild" and describes them as "an integral part of the natural system of public lands." But this vocabulary contradicts the management policies that have followed. The act gave the BLM jurisdiction to manage most of the wild horses on public lands in the West and to round up and remove excess horses, if needed, to maintain a balanced ecosystem. There was logic to this choice, given that many wild horses lived on land the BLM managed, and the agency had dealt with them out of necessity in the past.

But the overall task of the BLM, which grew out of the Grazing Service, is to balance the sustainable use of public lands in the West for a variety of purposes, including grazing, mining, and recreation. It is not primarily a wildlife agency. It deals in domestic livestock, and its business with wild horses had always been in reducing their competition for forage with domestic livestock. As you'll read, the agency has been criticized more recently for not using evidence-based wildlife management practices in its wild horse management program.

Following the act, the BLM established 219 Herd

Management Areas (HMAs), which are federal lands managed primarily, but not exclusively, for wild horses. Today there are 177 HMAs across ten states in the West. Only horses living in HMAs are subject to protections under the 1971 law. Numerous agencies, including the U.S. Forest Service, U.S. Fish and Wildlife Service (FWS), and National Park Service (NPS), as well as private organizations, such as the Corolla Wild Horse Fund and the Chincoteague Volunteer Fire Company, manage other wild horse populations across the country.

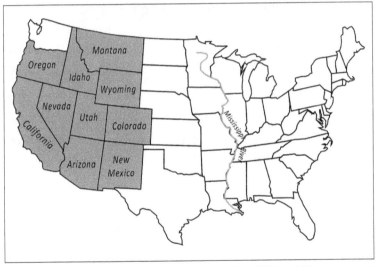

The ten states across which the BLM maintains 177 Herd Management Areas for wild horses.

Although the BLM is careful to use the term "wild" in reference to the wild horses it manages, as designated by the 1971 act, it is clear the federal government considers wild horses nonnative and sometimes feral, rather than native and wild, in its treatment of them. The BLM itself states: "The disappearance of the horse from the Western Hemisphere for ten thousand years supports the position that today's American wild horses should not be considered 'native.'"[530]

FWS, which manages a small number of wild horses on National Wildlife Refuges, acknowledges that the BLM refers to wild horses as "wild" because of the 1971 law. But FWS designates these animals as non-indigenous and feral,[531] and manages them as such, giving priority to plant and animal species that it designates as native.[532]

NPS considers wild horses nonnative and feral.[533] The agency manages a limited number of herds on federal park land for cultural value, including herds on Assateague Island National Seashore in Maryland, Cape Lookout National Seashore (Shackleford Banks) in North Carolina, Cumberland Island National Seashore in Georgia, and in Theodore Roosevelt National Park in North Dakota. But horses reside in numerous other NPS units, where the agency considers them trespass animals and works to remove them.

Despite inconsistent classifications across agencies in using the terms wild and feral, the federal government consistently labels wild horses nonnative, meaning species designated as native may receive priority on public lands, with some exceptions due to the 1971 act.

The 1971 act was a win for the advocates of America's wild horses. But between its questionable implementation and unforeseen herd population issues, protecting wild horses has presented significant challenges in the West.

39

STRUGGLES WITH WILD HORSE POPULATION CONTROL

It was almost unimaginable at the time the Wild Free-Roaming Horse and Burros Act of 1971 passed that America's twenty thousand remaining mustangs would be capable of much of a comeback, even with their new, protected status. Horse advocates had witnessed a steady decline in wild horse populations during their lifetimes, and the bill was really a last-ditch effort to recognize the horses' right to exist and to save a piece of the country's heritage. It failed to adequately plan, however, for what would happen when people stopped removing mustangs from the wild.

Now we understand that in an ecosystem with adequate forage for horses, they reproduce prolifically. For example, studies on Assateague Island show that if horse bands are left uninterrupted there, about half the mares will give birth each year.[534] The National Academy of Sciences estimates the wild horse population growth rate, which balances births with deaths, at 15-20 percent each year.[535] Those rates are unsustainable given the amount of land wild horses are allowed to occupy on HMAs.

Protecting and managing the wild horses on federal lands is a hugely challenging and controversial job. But few people, neither wild horse advocates nor ranchers, would argue the BLM has handled this task well. In fact, most would probably argue they have failed miserably.

Since the 1970s, the BLM's main wild horse management goal has been to reduce the ever-growing population, currently at an estimated 88,000 horses, down to 27,000, insisting it could get the population under control if it could just reach this number, called the Appropriate Management Level (AML), through roundups.[536]

To accomplish the AML, the agency has spent the last forty-five years engaged in an endless cycle of aerial roundups, the very practice Velma Johnston set out to eliminate, when they deem a population too large for the land it occupies or when ranchers pressure them to do so. These roundups flush horses out of isolated areas and scare them into corral traps.

Wild horse advocates argue this process is cruel and inhumane, separating family bands and resulting in injuries and deaths. Once removed from the wild, some of these horses are adopted. But adoptions never come close to matching the number of horses removed from public lands. Most of them, presently about fifty thousand, are sent to long-term holding facilities to spend the rest of their lives on the taxpayers' dime—to the tune of $50 million per

year.[537]

The problems with this approach to population management are as numerous as they are contentious. First, the BLM has neglected to use rigorous scientific methods to determine an accurate size of the wild horse population or to develop effective population management plans, like those used by other agencies that manage wildlife.[538] The magical AML number of 27,000 has no scientific basis; it is meaningless.

Second, removing members of a population can have an undesired effect: growth rates can actually increase for two different reasons. If a mare loses her foal, she can become more fertile than she would otherwise have been at that time, a process known as compensatory reproduction.[539] A study on Assateague Island showed that in a herd that underwent an annual roundup of foals, the annual pregnancy rate of mares was 85 percent.[540] In a herd that did not lose foals to roundups, the annual pregnancy rate of mares was 45 percent. The BLM also keeps populations lower than the limits of available food. If populations were allowed to increase to the point of needing to compete for food, they would naturally have fewer foals and their population would eventually balance itself with the available resources.[541] This management policy is controversial, however, as some animals could suffer and starve in the process.

Third, since the BLM does not manage either of the wild horse's natural predators, the wolf or the mountain lion, it does not consider them in its population control strategies. FWS or state governments manage these animals, depending on the location.[542] The BLM does not acknowledge the mountain lion's potential to help keep wild horse populations in check either, despite studies showing that in territories where mountain lion and horse territories overlap, mountain lions could consume up to two-thirds of the foals in a herd per year.[543] Admittedly, it seems unlikely these predators would solve the horse population issues as a whole, since Americans have spent so much energy trying to eradicate these very animals, either to protect their livestock or for the thrill of the hunt alone.

Wild mare and foal of the Pryor Mountains of Montana.

Fourth, roundups and long-term holding are extremely expensive and eat up most of the BLM's budget for the wild horse program, leaving little room to pursue more effective management options.[544]

While the BLM technically has authority to slaughter wild horses removed from the wild under the 1971 law, today's American public tends not to accept slaughter of horses as a viable option. In fact, one poll suggests 80 percent of Americans oppose the slaughtering of horses.[545] So the BLM has not adopted a policy of slaughter as a management tool. Since 2010, congressional budget appropriations have actually prevented the BLM from using slaughter for population control, despite the agency's recent budget requests to do so.[546] That does not mean it has not happened or is not still happening, though.

Throughout the 1980s and 1990s, scandals rocked the BLM's wild horse program. First there was a policy in the 1980s that waived the fee for anyone adopting more than one hundred horses at a time. Nearly twenty thousand horses were secretly sent to slaughter by the truckload through this program.[547]

In the 1990s, a whistleblower alerted authorities to another scandal. This time, BLM employees were accused of turning a profit under the table by conspiring with middlemen to funnel thousands of captured wild horses to slaughter.[548] Perhaps the most concerning part was how high the cover-up extended into the federal government; the Department of Justice pressured the prosecuting attorney to drop all charges.[549] Although no one was ever held legally responsible for the scheme or the cover-up, Associated Press journalist Martha Mendoza's investigative reporting publicly exposed the BLM's corruption—another major blow to the agency's reputation.[550]

Although the BLM has tightened its adoption policies since these scandals, questions still arise to this day as to whether middlemen are gaining access to wild horses for slaughter.[551] You would be hard-pressed to find someone with a lot of confidence in the BLM's ability to figure out how to effectively and ethically manage America's wild horse population.

But there is some good news, too: another method of population control has time and again shown promise in managing wild horse populations and reducing or preventing the need for roundups altogether: porcine zona pellucida (PZP), an immuno-contraceptive that prevents pregnancy in mares. It needs to be administered by dart annually, but it is highly effective and inexpensive. Unlike male and female sterilization, PZP does not have significant side effects and is easily reversible. The question is: will the BLM ever use PZP on a meaningful scale?

40

THE ORIGINS OF PZP

The origin story of using PZP to control wild horse populations shows just how much grit and failure scientific innovation requires, as well as how the resolution to one particular problem can sometimes be translated more broadly to other fields.

Shortly after the 1971 law went into effect, two BLM biologists tasked with managing the Pryor Mountain Wild Horse Range in Montana and Wyoming approached Jay Kirkpatrick, a wildlife reproductive physiologist who had just started teaching at Montana State University. They asked if he could figure out a way to keep wild horses from reproducing.[552]

Kirkpatrick agreed to try. He enlisted John Turner, a professor of physiology and endocrinology, as his research partner, and the team got to work trying to develop a form of fertility control for wild horses that was both effective and feasible to administer.

Up to that time, no one had published research papers on wild horse biology, so Kirkpatrick and Turner realized they had no foundation from which to launch their own studies.[553] Ultimately, their research would form the seminal body of work on the biology of America's wild horse population, but that meant it would take time and a lot of background research to develop and test a contraceptive strategy—more than twenty years.

The team first spent years observing wild horse behavior and trying to determine how to get blood samples for their research. By 1984, they were ready to test a hormone used to reduce fertility in stallions that seemed promising during their experiments out West.

However, they needed to find a smaller, more isolated population for field testing. Assateague Island seemed like the perfect place. Wild horses on the eastern barrier islands are managed by different agencies and organizations, not by the BLM. NPS manages the part of the Assateague herd that lives on the National Seashore in Maryland, while the Chincoteague Volunteer Fire Company manages the herd on the Chincoteague National Wildlife Refuge on the Virginia side of the island through a permit issued by FWS. NPS invited the team to test out fertility control on their herd, but with a catch: they wanted fertility control for the *mares*. NPS did not allow roundups of the herd either, so the researchers needed to figure out how to administer contraceptives by dart gun.

Kirkpatrick and Turner's first attempts, starting in 1986,

A wild horse grazes on the dunes of an eastern barrier island.

were utter failures. Figuring out darts that worked properly for the task was laborious, but the real problem lay in the fertility control itself. They started with a hormonal birth control similar to types effective in humans.

When they later tested the six mares they had darted for pregnancy, they anticipated less than half would be pregnant, since they had determined the birth rate for untreated wild mares on Assateague to be about 50 percent. But disaster struck: all six mares were pregnant![554] What the researchers thought would function as birth control actually turned out to work as a fertility enhancer in horses, spurring immediate ovulation.

The team went back to the drawing board and eventually found PZP, which had been isolated as a method of fertility control by Irwin Lui at the University of California-Davis. Rather than interfering with hormones, PZP works by triggering an immune response that prevents sperm from locking into an egg and fertilizing it.

Mare and foal seek shelter on the sand dunes from an incoming storm.

In 1988, the team darted twenty-six wild mares with PZP on Assateague, the first wild mammals of any sort to receive the treatment. When the team later tested the mares for pregnancy, the results were astounding: not one mare was pregnant. Further research confirmed they had found a highly effective form of fertility control for wild horses. By 1994, NPS had declared PZP a successful method for controlling the herd on Assateague.

Later studies demonstrated PZP to be effective on just

about any mammal with hooves, as well as elephants, hippos, and bats—a major victory for the management of wildlife populations.[555]

PZP comes with its own challenges, though. Mares must be located, properly identified, and darted annually, though newer formulations of PZP in the works may stretch effectiveness to several years. People must be trained to do the painstaking work. And management agencies must be convinced to trade their traditional methods of population control for something new.

But a huge benefit of PZP, aside from being a more humane and effective way to control populations than roundups and removals, is that it is cheap: $30 per year per horse, compared to $1,829 per year per horse to keep it in short-term holding immediately following roundup, and $664 per year per horse for the lifetime of the horse to keep it in long-term holding.[556]

Numerous agencies managing herds in the East now successfully use PZP for population control. Many volunteer groups in the West now dart horses to try to reduce roundups. The BLM does use PZP in a few herds, including in the Pryor Mountains, but spends less than 1 percent of its $80 million budget on this method of fertility control.[557] In 2018, that percentage amounted to just 702 mares treated with PZP.[558]

Time and again, scientific agencies, including the National Academy of Sciences, have recommended the BLM change its management program to incorporate PZP on a large scale to reduce roundups and long-term holding.[559] Time and again, the BLM, caught in a cycle of spending the majority of its budget on roundups and holding, does not make appreciable changes to use PZP in a meaningful way.

Wild foal.

As of the writing of this book, the Trump administration just backed down, due to political pressure, from an attempt to put in place a slaughter policy for horses removed from the wild and not adopted, which it had proposed for several years. Instead of pursuing PZP as a viable and relatively inexpensive solution to population management, however, the administration is considering pursing the surgical sterilization of wild mares by removing their ovaries—a method not supported by the National Academy of Sciences or veterinary organizations.[560]

In 2019, after three years of negotiations, a consortium of ranching, hunting, and animal advocate groups, including the Humane Society of the United States and Return to Freedom Wild Horse Conservation, presented a ten-year wild horse population control plan to the BLM and Congress.[561] This plan lays out a strategy to reduce herd size to near the AML, which we've already established as a baseless number, within a decade. It includes significantly increased roundups and removals (twenty thousand removals annually for the next three years, then five thousand to ten thousand for the following seven years); PZP fertility treatment for 90 percent of the remaining mares; the increased use of less expensive, private pastures for long-term storage; and increased adoptions.

The plan, criticized by some other animal welfare advocates as a major win for the livestock industry, would add an additional $50 million to the BLM's annual wild horse and burro program budget, which is already around $80 million.[562] In December 2019, Congress allocated a $21 million increase to the program's budget, pending the submission of a five-year plan by the BLM that includes fertility control measures and the removal of horses to long-term holding facilities.[563] The budget allocation is largely seen as a nod to the proposed wild horse population control plan.

Little did Velma Johnston and her fellow advocates

know that by demanding the protection of America's last remaining wild horses and the cultural heritage these horses represented, they launched a political battle that would rage into the next century. Despite the 1971 bill that called wild horses "wild," the federal government has largely considered them nonnative and feral in their treatment, as evidenced by the BLM's management practices.

There is no end in sight to the controversy of how to consider and handle, both scientifically and practically, America's wild horses. Questions continue to brew. Despite the 1971 law protecting them, do wild horses actually have the right to exist in the U.S.? What price is worth preserving the cultural heritage these horses represent? And how long are we willing to keep paying it?

41

MODERN THOROUGHBRED RACING

M odern Thoroughbred racing is a sport mired in controversy. It is also a $25-billion-per-year industry that involves about 1.2 million horses at any given time, and is heavily protected by powerful gambling lobbyists.[564]

About ten race horses die per week on U.S. tracks, a figure two and a half to five times greater than the rates outside the U.S.[565] This discrepancy may be due to the fact that, unlike most countries with sizeable racing industries, the U.S. lacks regulation at a federal level. Instead, the thirty-eight states that authorize racing determine their own regulations, meaning rules that govern the sport and treatment of its horses change every time you cross a state line. The inconsistency and lack of oversight give cheaters and exploiters leeway to act without much penalty.

Performance-enhancing drugs are regularly used and abused with little to no regulation or consequence.[566] Not only do these drugs threaten the integrity of the sport, just as doping does for human sports, but they are also dangerous to the horses. Heavy pain medications can conceal injuries both from a vet certifying a horse is healthy enough to race and from the horse itself, leaving the animal at risk of running too hard on an injury and causing further injury or death. A broken leg is enough to kill a horse. Many racehorse vets also actively participate in over-drugging and providing banned substances; selling and administering drugs actually make up most of their incomes.[567]

Just one example: virtually all Thoroughbred racehorses in the U.S. are administered Lasix (furosemide), a diuretic, before racing.[568] This race-day practice is banned in most other countries. Some horses are susceptible to bleeding in their lungs if they run too hard, and Lasix can cut down bleeding by about 50 percent for some of these horses. But this lung condition only causes significant issues in less than 10 percent of horses. So why do 95 percent of them receive Lasix before racing? Because it causes them to urinate so much they immediately drop twenty to thirty pounds, boosting their speed by three to five lengths. The practice has become so widespread that horses not drugged with Lasix are no longer competitive.

Accusations of animal cruelty in the industry abound, including overworking horses, forcing them to run on dangerous injuries, using tools to administer electrical shocks to make horses run faster, and running them on tracks when the surfaces are likely to cause injury.[569]

The deaths of twenty-three horses, mainly from limb injuries, between late December 2018 and early 2019 on one track alone, the famed Santa Anita Park in California, caused a stir and caught the public's attention. The track shut down for about a month, and the Los Angeles County District Attorney's Office launched an investigation into the spike.[570]

Santa Anita, owned by the Stronach Group, tolerated trainers who had been cited for performance-enhancing drugs.[571] Critics have accused the company of putting profit over equine safety, pushing races in unsafe conditions.

Although the backlash from these deaths was a wakeup call to the industry, the deaths did not stop there. By June 2019, the total had increased to twenty-nine horse deaths at Santa Anita since December 2018. That month, Governor Gavin Newsom of California called for a halt to racing at Santa Anita until independent veterinarians could evaluate the horses for fitness.[572] Track owners refused to close it but agreed to increased evaluation of the horses.[573]

The deaths continued. In December 2019, the Los Angeles County District Attorney's Office investigation revealed that between July 1, 2018, and November 30, 2019, fifty-six horses had died at Santa Anita.[574] The investigation stated: "there was insufficient evidence to prove criminal animal cruelty or other unlawful conduct under California law." But it listed numerous "areas of concerns and possible factors that may have contributed to the deaths," including medications found in many of the horses (which were not illegal given California's regulations); the track conditions on which the horses were running (some witnesses thought the "sealed track" in rainy conditions may have increased the risk of limb injuries); the possibility of running injured horses (necropsies revealed preexisting conditions in some of the dead horses, but the investigation could not prove trainers or veterinarians had knowledge of these conditions and raced the horses anyway); and undue pressure to race despite track conditions (the investigation did not find evidence of this, however). The report called the investigation "ongoing" and recommended the "California horse racing industry and regulators coordinate their efforts to formalize strategic safety plans aimed at reducing equine fatalities."

Santa Anita has instituted some recommended safety changes since the investigation began. In the first three weeks of 2020, however, five more horses died on the track.[575]

Meanwhile, in September 2019, news broke about potential drug use in a high-profile racehorse. Justify, the 2018 winner of racing's prestigious Triple Crown, had failed a drug test for a banned substance, scopolamine, weeks before the Kentucky Derby, the first of the three-race series.[576] Scopolamine may influence a horse's breathing and heart rate. Rather than filing a public complaint and disqualifying Justify as protocol dictated, the California Horse Racing Board sat on the information for a month. Then it privately decided to reduce the punishment for scopolamine, considered a performance enhancer, especially in the dosage the test revealed, to a fine and a possible suspension. (Some veterinarians have pointed out it is possible for a horse to get a positive scopolamine test following the accidental ingestion of jimson weed.[577]) Citing environmental contamination, the board also dropped the case against Justify, whose breeding rights had already been sold for $60 million. Regardless of whether Justify had inadvertently ingested jimson weed or been purposely drugged, the case was yet another public relations nightmare for the industry.

In March 2020, twenty-seven racehorse trainers, veterinarians, and drug distributors were indicted on federal charges relating to an elaborate horse-doping scheme.[578] The four indictments accused individuals of participating in manufacturing, mislabeling, distributing, or administering banned drugs in an effort to cheat in high-profile races for large profits. Among those charged was Jason Servis, trainer of Maximum Security, the horse that finished first in the Kentucky Derby in May 2019, but was later disqualified, and went on to win a $10 million race purse just before the indictments hit.

The upticks in horse deaths at Santa Anita and ongoing scandals within the industry come at a time when Thoroughbred racing is already stumbling.[579] The number of races has been cut in half since 1990, and on-track betting has declined by about a half since 2001. Public interest is waning. Newly legalized forms of betting could also move money away from the tracks.

These circumstances together have put Thoroughbred racing on notice. In March 2019, U.S. lawmakers introduced the Horseracing Integrity Act in an effort to save the industry. Representative Paul Tonko of New York, who co-sponsored the bill, said, "This will make horse racing safer for our equine athletes and jockeys while increasing confidence in the sport among the trainers, owners, horseplayers, and horse racing fans alike."[580]

The act would set uniform standards for drug use across the racing industry; create an independent anti-doping organization to monitor drug use and enforce penalties for abuse; ban all drugs within twenty-four hours of racing; and increase safety regulations for horses, jockeys, and drivers.[581] The bill, yet to pass as this book goes to press, has garnered the support of numerous racing and veterinary organizations that recognize the industry has work to do to stay afloat in an era of continuous scandal and dwindling fans.

42

THE HORSE INDUSTRY TODAY

The horse may have lost its position as the country's chief power source more than a hundred years ago, but it still has an enormous influence on the U.S. economy: $50 billion per year in direct economic impacts, including about one million jobs.[582] When you add indirect impacts that reach other sectors, the economic impact jumps to $122 billion per year—a greater influence on the economy than the motion picture industry.[583]

About 7.2 million horses live in the U.S., down from the country's peak of around twenty-five million horses in 1900.[584] Whereas most horses in 1900 were work animals, nearly half the horses in the country today are used for recreation.[585] The rest are split between showing, racing, and other work.

It should be no surprise that three states with historically significant horse populations have the largest horse populations today: Texas, California, and Florida.[586] And it should also be no surprise that the versatile, all-purpose American Quarter Horse is still the country's most popular breed.[587]

Throughout the development of the U.S., men likely owned most of the horses in the country. Today, in a significant demographic shift, the majority of the estimated two million horse owners in the U.S. are women.[588]

Keeping horses today is extremely expensive. Given that most communities do not have stables on every corner, most people do not live on farms, and many horses are not earning their livings, owning them has largely become, once again, a privilege of the elite.

Recreational uses for horses include many of the same activities that have been happening for hundreds of years in the U.S.: pleasure riding, fox hunting, and polo.

Showing today involves a wide variety of activities and styles. English shows include a number of categories, such as equitation, jumping, and dressage—a competitive art that began to surge in popularity in the U.S. in the 1970s in which horse and rider perform precision movements that require high skill and intense training. Western shows and rodeos often include sports that demonstrate the skills cowgirls and cowboys use to work cattle, such as roping, cutting, and barrel racing. Western pleasure riding evaluates the horse on its ability to perform slow, controlled movements. Participants of equestrian vaulting perform gymnastics on horseback.

Although Thoroughbred racing is the highest-profile

form of horse racing today, harness racing, steeplechases, and endurance (long-distance) events are popular, as well.

Horses are well past their heyday as energy sources but still hold many of the jobs they always have. They work in a variety of ways throughout communities, including for riding lessons, as police mounts, and as agricultural workers.

As we've discussed, a less-quantifiable characteristic of horses is the social bond they are able to form with humans, which speaks to why so many people in the U.S. still own horses today. Humans have likely been using horses for therapeutic purposes for thousands of years, but it is only more recently that this process has been formalized through the introduction of equine-assisted activities and therapies (EAAT). EAAT encompasses a broad spectrum of equine activities used to aid in the treatment of mental health issues, behavioral disorders, and some gross motor challenges.

In 1969, the organization now known as the Professional Association of Therapeutic Horsemanship International (PATH) brought the practice of EAAT to the U.S. These programs now exist all over the country for at-risk children and adults, as well as for survivors of abuse and people with autism, ADHD, eating disorders, cerebral palsy, sensory delays, and other physical, emotional, or developmental issues. Some EAAT activities focus on grooming and caring for horses, while others emphasize riding or carriage driving.

The intensity of the human-horse bond has the power to influence people's lives for the better. Some equine programs work on building relationships, self-confidence, and communications skills. For example, one organization in North Carolina, CORRAL Riding Academy, has combined horse therapy, academic tutoring, and mentorship to help girls in high-risk situations heal and put themselves on track for academic success. Eighty-eight percent of the girls who complete their program go to college.[589]

Many programs provide EAAT to military veterans who have sustained physical injuries or developed mental health issues related to their service.[590] These programs pair veterans with horses for groundwork, grooming, riding, and horsemanship lessons in an effort to help veterans heal.

Programs associated with the BLM pair prison inmates with mustangs removed from the wild. The inmates saddle train the horses to prepare them for adoption, but the idea is that the gentling works both ways.[591] Learning social and vocational skills, as well as how to take responsibility for and have pride in their work can lead to lower rates of recidivism.[592]

Hippotherapy, a form of EAAT that arrived in the U.S. in the 1900s, incorporates horses into physical, occupational, and speech therapies. For example, children with cerebral palsy have had success improving gross motor skills from sitting on moving horses.[593]

Formalized EAAT programs have not existed or been

standardized long enough to accumulate much data evaluating their effectiveness. In fact, studies really only began to delve into EAAT in the early 2000s. Although most case studies to date have used small sample sizes and are difficult to compare across programs, many demonstrate that EAAT has potential to contribute to therapy for numerous conditions and circumstances. Anecdotal evidence of positive outcomes, such as from CORRAL Riding Academy and BLM programs with inmates, supports these studies. With the increase in EAAT programs around the country, it is likely that larger-scale studies will soon offer more information about the strengths and limitations of equine therapy.

But anyone who has worked with horses understands their loyalty and the social bonds they form with humans. These characteristics are likely the reasons so many Americans continue to keep horses for the pleasure of it, long after electricity and combustion engines replaced them in most industries.

43

CONCLUSION

As a critical player in U.S. history, the horse shaped the country in profound ways. Through its strength, versatility, and ever-presence, the horse served as one of the most powerful influences on the development of the U.S., a common thread binding together seemingly disparate events and offering a unique perspective on how this country came to be.

Horses acted as weapons of domination, first for the Spanish and then for the indigenous peoples who learned to wield Spanish horses so rapidly and skillfully for warfare that they kept Europeans from much expansion into the Great Plains for two centuries. The U.S. government failed to subdue these Horse Nations until it understood the only way to do so was to remove their horses. In both the Revolutionary and Civil Wars, horses helped to shift the balance to the winning side.

Horses defined numerous cultures. They transformed

the Comanche and other plains peoples from semi-nomadic hunter-gatherers into efficient hunters and fully nomadic warriors, whose cultures revolved almost entirely around the horse. Western cowboys also established nomadic customs, living in the saddle as they moved cattle across the plains. Spanish colonists who planted missions up the coast of California developed a horse-centered culture, too. Children rode from an early age, and artistic endeavors hinged on horses. On the opposite coast, early colonists in Virginia formed their own culture around horse breeding and racing.

Wild horses gather by the water to cool off on the Outer Banks.

Horses served as status symbols and currency. They acted as entertainers, moneymakers, and emblems of liberty. They powered the developing nation by hauling vehicles, towing barges, pulling train engines, and assisting in construction and agriculture.

Horses fueled literature and legend. They have been firmly embedded in our national imagination throughout the development of the country. Lore of the pacing white stallion and the romanticized image of the cowboy endure today. Equine literature also gives us insight into the intensity of the horse-human bond.

Horses empowered minorities and women. They elevated the status of black jockeys, trainers, and cowboys above that of other blacks, and provided a means for women to assert themselves in the male-dominated professions of ranching and rodeos.

Horses have incited bitter controversy, from animal welfare debates, to battles over wild horse populations, to the struggles of today's Thoroughbred racing industry.

Horses served as loyal companions. Most members of the Horse Nations, colonists, explorers, homesteaders, cowboys, soldiers, farmers, and miners of the early U.S. likely shared something in common: time spent on the ever-shifting frontier with nothing but their loyal mounts for company. Today, technology has rendered horses obsolete in many industries, but that has not stopped millions of Americans from seeking out their company, whether through casual riding, admiration of the remaining wild horses, or equine therapy. The power of that bond endures.

TIMELINE

66 MYA—Dinosaurs become extinct

55 MYA—Dawn horse (*Hyracotherium/Eohippus*) emerges

6-3 MYA—The modern horse genus, *Equus*, emerges

2.6 MYA—Ice Age begins

1.7 MYA—The modern horse species, *Equus caballus*, emerges in North America

15,000 YA—First humans migrate across land bridge from Asia, if not earlier

12,000-10,000 YA—Ice Age ends

11,000-8,000 YA—*Equus* becomes extinct in North America (Quaternary Extinction)

1493—Columbus brings first modern horses to the Americas on his second voyage

1519—Cortés brings first modern horses to the mainland of North America

1521—Ponce de León fails to colonize Florida

1526—Ayllón fails to colonize South Carolina and Georgia

1527—Narváez fails to colonize Florida

1539—De Soto fails to colonize the Southeast

1540—Coronado traipses through New Mexico

1584—Britain's first exploratory voyage to the Outer Banks of North Carolina

1585—Britain's first colonization attempt of the Outer Banks

1587—Britain's second colonization attempt of the Outer Banks (the Lost Colony)

1590—Britain's White returns unsuccessfully to look for the Lost Colony

1598—Oñate starts settlement around Santa Fe, New Mexico

1500s-1600s—Legends of horses swimming ashore from sinking Spanish ships onto eastern barrier islands

1607—British establish first permanent colony in Jamestown, Virginia

1615—Spanish have established 20 missions across Florida and southern Georgia

Early 1600s—Horse spreads from Florida and Georgia to Cherokee, Chickasaw, Choctaw, Creek, and Seminole nations

1619—Jamestown colonists purchase 20-30 enslaved Africans, launching the African slave trade in what would become the U.S.

1621—Spanish allow Pueblo indentured servants to ride horses if they convert to Catholicism

1624—Dutch settle in New York and import draft horses from the Netherlands

1625—The small number of British horses imported to Virginia have died

1632—Only one documented horse living in the Plymouth Colony in Massachusetts

1640s—Battles between British settlers and Native Americans scatter horses to the wild; these bands spread through the Blue Ridge Mountains

1640s—Some Navajo and Apache chiefs learn to ride

1649—Only 200 documented domestic horses in Virginia

1650—Spanish have established 72 missions across Florida and southern Georgia

1650s—Reports of Apache stealing horses from Spanish

1650s—Settlers begin to pasture horses on the Outer Banks

1670—British fencing tax results in more colonists pasturing horses on the Outer Banks

1680—Pueblo Revolt in New Mexico; Great Horse Dispersal

1680—Comanche acquire horses

Late 1600s—British colonists pasture their horses on Assateague Island in Maryland and Virginia

Late 1600s—Narragansett Pacer, considered the first truly American breed, emerges in Rhode Island

1700—All Native American nations in Texas have acquired horses

Early 1700s—Raids by British settlers shut down Spanish missions in Georgia and Florida

1730—Nez Perce acquire horses

1730s—Virginians begin to import Thoroughbreds from England for racing

1732—Georgia, the last of the original 13 colonies, is established

1750—All native peoples of the plains have acquired horses as far north as Canada

1750—Comanche have established the large territory of Comancheria across the southern Great Plains

Mid-1700s—Chickasaw begin to acquire horses from the Texas plains for trading

1756—English Thoroughbred Janus arrives in Virginia

1769—Spanish begin to colonize California; 300,000 native peoples live there

1775—Revolutionary War begins

1775—Betsy Dowdy's legendary ride

1775—Paul Revere's legendary ride

1776—U.S. Declaration of Independence

1777—Sybil Ludington's legendary ride

1783—Treaty of Paris ends the Revolutionary War, drawing the western borders of the U.S. at the Mississippi River

1788—First documentation of wild horses on Cumberland Island, Georgia

1780s or 1790s—The horse Justin Morgan is born

1793—Eli Whitney invents the cotton gin; popularity of the mule for plantation work explodes

1795—Philadelphia and Lancaster turnpike opens

1800—Narragansett Pacer has mostly disappeared

1800—California has around 24,000 horses

1803—Louisiana Purchase doubles the size of the U.S.

1804—Lewis and Clark Expedition sets out to explore the new U.S. territory

1805—Californians have more horses than they can use and begin to slaughter them

1817—Erie Canal constructions begins

1821—Mexico wins independence from Spain; invites

Americans to settle in Texas to act as buffer against Comanche

1821—Santa Fe Trail carved out by pack train of mules

1829—First omnibus introduced in New York City

1830—Indian Removal Act begins the 20-year-long Trail of Tears

1832—Washington Irving publishes *A Tour of the Prairies*

1833—Mexico dismantles mission system and distributes mission land in California and the Southwest to Mexican settlers

1836—First wagon train heads west from Missouri to Idaho on what would become the Oregon Trail

1836—Texas becomes an independent republic

1837—Financial crisis

1843—Hays and his Texas Rangers begin to carry the new Colt revolver to fight the Comanche

1845—U.S. acquires Texas

1845—The term manifest destiny is coined

1848—California Gold Rush begins

Mid-1800s—Quarter Horse falls out of fashion for racing in the East

Mid-1800s—Several million wild horses roam the West during their prime

1850s—Streetcar replaces omnibus

1851—Indian Appropriations Act designates fund for U.S. government to move Native Americans of the West onto reservations in Oklahoma

1852—California Gold Rush peaks

1855—400 Comanche agree to move to a reservation

1855—Some Nez Perce sign Treaty of Walla Walla

1857—Invention of machine to mass-produce horseshoes

1860—Prospector finds gold on Nez Perce territory; white fortune-seekers overrun Nez Perce land

1860—Pony Express launches

1861—First telegraph line connects East and West, closing the Pony Express after 18 months of operation

1861—Civil War begins

1862—The Homestead Act established; the Quarter Horse travels West with onslaught of settlers

1863—Grierson's cavalry raid

1863—The Saratoga Race Course opens

1865—Civil War ends

1865—Reconstruction Era begins

1865—40 Acres and a Mule begins and is quickly reversed

1866—Transcontinental railroad arrives in Kansas, making the western cattle industry profitable and launching the heyday of the nomadic cowboy

1867—Several hundred Comanche sign Medicine Lodge Treaty and agree to move to reservation; now 1,000 Comanche live on reservation

1868—Americans kill 31 million buffalo in the following 13 years

1868—14th Amendment ratified, granting citizenship and equal protection to all persons born or naturalized in the U.S., including the formerly enslaved

Late 1800s—One-quarter of cowboys in the West are black

Late 1800s—Second Industrial Revolution begins

1870—Native American population of California down to just 30,000

1872—The Great Epizootic sweeps through horses in almost every major city in the U.S. and Canada in 50 weeks

1873—Invention of barbed wire, which fences off the Great Plains

1875—Last of Comanche forced to reservation

1876—Battle of Little Bighorn

1877—Nez Perce War

Late 1870s—U.S. government moves the last of the Horse Nations to reservations

1877—Reconstruction Era ends

1877—Anna Sewell publishes *Black Beauty*

1880s—Bounty hunters earn $25 per wild stallion scalp

1885—August Belmont moves Thoroughbred breeding operations to Kentucky

1886-1887—Cattle industry crashes due to blizzard conditions and barbed wire

1890—American Humane Education Society publishes and distributes Sewell's *Black Beauty* in the U.S.

1891— Oldest horse breed association in the U.S. established, the American Saddlebred Horse Association

Late 1800s—Wild West Shows becomes popular

Late 1800s—First accredited veterinary schools established in U.S., primarily to care for horses

1900—Theodore Roosevelt coins the term "cowgirl"

1902—Owen Wister publishes *The Virginian*

1902—Electricity fuels almost all streetcars

1908—Henry Ford releases assembly-line-produced Model T

Early 1900s—Segregationist laws ban black jockeys from racing

Early 1900s—Second Industrial Revolution ends

1910—Peak of non-farm use of horse

1912—Number of cars surpass number of horses on road in New York City

1914—WWI begins

1918—WWI ends

1920s—P.M. Chappel introduces canned dog food made from slaughtered horses

1929—Great Depression begins

1938—Piloted roundups of wild horses begin

1939—End of the Great Depression

1939—WWII begins; Marsh Tackies patrol South Carolina beaches throughout the war

1941—Great Louisiana Maneuvers; Army decides to leave most horses at home for WWII

1942—26th Cavalry's last charge in the Philippines

1945—Army rescues Lipizzaners and Arabians from Nazis

1945—WWII ends

1945—Marguerite Henry publishes *Justin Morgan Had a Horse*

1947—Marguerite Henry publishes *Misty of Chincoteague*

1950s—An estimated 20,000 wild horses remain in the West

1959—Congress passes first federal Wild Horse Annie Act

1966—Marguerite Henry publishes *Mustang, Wild Spirit of the West*

1968—Establishment of Pryor Mountain Wild Horse Range on Montana-Wyoming border, the first of its kind

1969—PATH brings equine-assisted activities and therapies (EAAT) to U.S.

1970s—Dressage introduced in U.S.

1971—The Wild Free-Roaming Horses and Burros Act passes, granting wild horses on federal lands a measure of protection

1980s—BLM implicated in scandals that resulted in the slaughter of wild horses (through 1990s)

1988—PZP used successfully in wild horses for the first time (on Assateague Island)

1990s—Hippotherapy arrives in U.S.

2001—Green Berets enter mounted combat in Afghanistan

2019—Horse deaths at Santa Anita capture national attention

2019—U.S. lawmakers introduce the Horseracing Integrity Act (still pending)

2019—Consortium of organizations present 10-year wild horse population control plan to the BLM and Congress

2020—27 racehorse trainers, veterinarians, and drug makers indicted on federal charges of horse doping

NOTES

[1] (Story of Emtech) Associated Press, "Emtech Breaks Down at Santa Anita, Becomes 32nd Horse to Die at Track since December," *USA Today*, September 28, 2010.
https://www.usatoday.com/story/sports/horseracing/2019/09/28/santa-anita-emtech-32nd-horse-die-track-december/3809534002/
[2] (Facts in paragraph.) Dan Ross, "Safety and Welfare with Breeder's Cup in Sight," *Thoroughbred Daily News*, October 22, 2019.
https://www.thoroughbreddailynews.com/santa-anita-safety-and-welfare-with-breeders-cup-in-sight/
[3] Washington Irving, *A Tour on the Prairies,* 1835. Cited in J. Frank Dobie, *The Mustangs* (Lincoln: University of Nebraska Press, 2005), 144. (Original work published in 1934.)
[4] Ann Norton Greene, *Horses at Work: Harnessing Power in Industrial America* (Cambridge: Harvard University Press, 2008), 202, 242-243.
[5] John Clayton, *The Cowboy Girl: The Life of Caroline Lockhart* (Lincoln: University of Nebraska Press, 2007), 41.
[6] Judith Dutson, *Storey's Illustrated Guide to 96 Horse Breeds of North America* (North Adams: Storey Publishing, 2005), 24.
[7] David Philipps, *Wild Horse Country* (New York: W.W. Norton & Company, 2017), 33.
[8] Wendy Williams, *The Horse: The Epic History of Our Noble Companion* (New York: Scientific American, 2015), 55.
[9] Williams, *The Horse*, 118.
[10] (Facts in paragraph.) Philipps, *Wild Horse Country*, 23.
[11] Ibid, 25.
[12] Ibid, 27.
[13] Philipps, 27.
[14] George Gaylord Simpson, *Horses* (New York: The American Museum of Natural History, 1961), 179. (Original work published in 1951.)
[15] Philipps, *Wild Horse Country,* 24.
[16] Jay Kirkpatrick and Patricia Fazio, "The Surprising History of America's Wild Horses," *Live Science*, July 24, 2008. https://www.livescience.com/9589-surprising-history-america-wild-horses.html
[17] Ibid.
[18] Williams, *The Horse*, 131-132.

[19] Ibid, 133.

[20] Craig Childs, *Atlas of a Lost World: Travels in Ice Age America* (New York: Vintage Books, 2019), xiv.

[21] Ibid, 180.

[22] Ibid, 197.

[23] Williams, *The Horse*, 128.

[24] Kirkpatrick and Fazio, *"The Surprising History of America's Wild Horses."*

[25] Williams, *The Horse*, 178.

[26] Melanie Pruvost, Rebecca Bellone, Norbert Benecke, Edson Sandoval-Castellanos, Michael Cieslak, Tatyana Kuznetsova, Arturo Morales-Muñiz, Terry O'Connor, Monika Reissmann, Michael Hofreiter, and Arne Ludwig, "Genotypes of predomestic horses match phenotypes painted in Paleolithic works of cave art," *PNAS*, 108, 46 (November 15, 2011): 18626-18630. https://doi.org/10.1073/pnas.1108982108

[27] Ibid.

[28] (Facts in paragraph.) Ibid.

[29] (Facts in paragraph.) Anna Linderholm and Gregor Larsen, "The role of humans in facilitating and sustaining coat colour variation in domestic animals," *Seminars in Cell and Developmental Biology*, 24, 6–7 (June–July 2013): 587-593. https://www.sciencedirect.com/science/article/pii/S1084952113000517

[30] Hope Ryden, *America's Last Wild Horses* (Guilford: The Lyons Press, 1970), 26.

[31] (Facts in paragraph.) Ibid.

[32] Dobie, *The Mustangs*, 4.

[33] Ryden, *America's Last Wild Horses*, 24.

[34] (Facts in paragraph.) Dobie, *The Mustangs*, 22.

[35] Bernal Díaz del Castillo, *The Conquest of New Spain*. Cited in Dobie, *The Mustangs*, 25.

[36] Ibid, 21.

[37] (Facts in paragraph.) Philipps, *Wild Horse Country*, 39.

[38] (Facts in paragraph.) S.C. Gwynne, *Empire of the Summer Moon: Quanah Parker and the Rise and Fall of the Comanches, the Most Powerful Indian Tribe in American History* (New York: Scribner, 2010), 54.

[39] (Facts in paragraph.) Alexander Koch, Chris Brierley, Mark M. Maslin, Simon L. Lewis, "Earth Systems Impacts of the European Arrival and Great Dying in the Americas after 1492," *Quaternary Science Reviews*, 207 (March 1, 2019): 13-36. https://www.sciencedirect.com/science/article/pii/S0277379118307261#

[40] (Including the following paragraph.) Philipps, *Wild Horse Country*, 40.

[41] Ibid, 42.

[42] Ibid.

[43] Ibid, 45.

[44] Ibid, 42.

[45] (Dates tribes received horses.) Gwynne, *Empire of the Summer Moon*, 31.

[46] Haines, *Horses in America*, 68.

[47] (Facts in paragraph.) Dobie, *The Mustangs*, 48.

[48] Gwynne, *Empire of the Summer Moon*, 33.

[49] Dobie, *The Mustangs*, 73.

[50] Ibid, 79.

[51] Gwynne, *Empire of the Summer Moon*, 28.

[52] Ibid, 104.

[53] Ibid, 109.

[54] Ibid, 100.

[55] Ibid, 104.

[56] Philipps, *Wild Horse Country*, 4.

[57] A. S. Leopold, S. A. Cain, C. M. Cottam, I. N. Gabrielson, T. L. Kimball. *Wildlife Management in the National Parks,* March 4, 1963. Accessed January 4, 2020. https://www.nps.gov/parkhistory/online_books/admin_policies/policy4-leopold.htm

[58] Kirkpatrick and Fazio, *The Surprising History of America's Wild Horses.*

[59] Philipps, *Wild Horse Country*, 33.

[60] Kirkpatrick and Fazio, *The Surprising History of America's Wild Horses.*

[61] David Powell's doctoral dissertation research, cited in Frydenborg, *Wild Horse Scientists*, 40.

[62] Williams, *The Horse*, 26-27.

[63] (Facts in next four paragraphs.) Gwynne, *Empire of the Summer Moon*, 27-28.

[64] Ibid, 33.

[65] Philipps, *Wild Horse Country*, 52.

[66] Ryden, *America's Last Wild Horses*, 167.

[67] Gwynne, *Empire of the Summer Moon*, 35.

[68] (Facts in paragraph.) Ibid, 59.

[69] (Facts in paragraph.) Ibid, 33-34.

[70] Dobie, *The Mustangs*, 46.

[71] (Including facts in following paragraph.) Gwynne *Empire of the Summer Moon*, 32-34.

[72] Ibid, 32.

[73] Ibid, 61.

[74] (Facts in remainder of paragraph and the following.) Dobie, *The Mustangs*,

64-68.

[75] (Facts in paragraph.) Ibid, 84.

[76] Gwynne, *Empire of the Summer Moon*, 40.

[77] Kent Nerburn, *Chief Joseph & the Flight of the Nez Perce: The Untold Story of an American Tragedy* (New York: HarperSanFrancisco, 2006), 8.

[78] Haines, *Horses in America*, 77.

[79] Nerburn, *Chief Joseph & the Flight of the Nez Perce*, 8.

[80] Haines, *Horses in America*, 77.

[81] Nerburn, *Chief Joseph & the Flight of the Nez Perce*, 9.

[82] Haines, *Horses in America*, 77.

[83] Ibid, 97.

[84] Dobie, *The Mustangs*, 56.

[85] Pruvost et al., "Genotypes of predomestic horses match phenotypes painted in Paleolithic works of cave art."

[86] (Dates for France, Greece, Egypt, Italy, and Austria.) Dutson, *96 Horse Breeds of North America*, 74. (Dates for China and Persia.) Ryden, *America's Last Wild Horses*, 95.

[87] (Facts in paragraph.) Nerburn, *Chief Joseph & the Flight of the Nez Perce*, 99.

[88] Ryden, *America's Last Wild Horses*, 96.

[89] Nerburn, *Chief Joseph & the Flight of the Nez Perce*, 9.

[90] (Facts in paragraph.) Haines, *Horses in America*, 78.

[91] Nerburn, *Chief Joseph & the Flight of the Nez Perce*, 9.

[92] Ryden, *America's Last Wild Horses*, 97.

[93] (Facts in paragraph.) Philipps, *Wild Horse Country*, 40.

[94] Haines, *Horses in America*, 34.

[95] Ibid.

[96] (Facts about Zemourri.) Sam Haselby, "Muslims of Early America," *Aeon*, May 20, 2019. https://aeon.co/essays/muslims-lived-in-america-before-protestantism-even-existed

[97] (Facts in paragraph.) Dobie, *The Mustangs*, 29-30.

[98] Haines, *Horses in America*, 54.

[99] (Facts in paragraph.) Dobie, *The Mustangs*, 32.

[100] Claggett, Stephen R., "North Carolina's First Colonists: 12,000 Years Before Roanoke," *The Ligature*, 1986. Revised March 25, 1996. https://archaeology.ncdcr.gov/articles/north-carolinas-first-colonists-12000-years-roanoke

[101] Bonnie U. Gruenberg, *The Hoofprints Guide to The Wild Horses of Corolla* (Strasburg: Quagga Press, 2015), 14.

[102] Ibid, 16.

[103] Ibid.

[104] Ibid, 18.

[105] Ibid.

[106] David Stick, *The Outer Banks of North Carolina* (Chapel Hill: The University of North Carolina Press, 1958), 21.

[107] Gruenberg, *The Hoofprints Guide to The Wild Horses of Corolla*, 18.

[108] Haines, *Horses in America*, 54.

[109] Ibid.

[110] Gwynne, *Empire of the Summer Moon*, 55.

[111] Dutson, *96 Horse Breeds of North America*, 16.

[112] Haines, *Horses in America*, 54.

[113] Ibid.

[114] (Facts in paragraph.) Ibid.

[115] Dutson, *96 Horse Breeds of North America*, 16.

[116] (Facts in paragraph) Haines, 93.

[117] (Facts in paragraph.) Ibid, 54.

[118] Ibid, 65.

[119] (Facts in paragraph.) Nikole Hannah-Jones, "Our democracy's ideals were false when they were written. Black Americans have fought to make them true," *The New York Times Magazine*, August 14, 2019. https://www.nytimes.com/interactive/2019/08/14/magazine/black-history-american-democracy.html

[120] (Facts in paragraph.) Haines, *Horses in America*, 57.

[121] (Facts in paragraph.) Julie A. Campbell, *The Horse in Virginia: An Illustrated History* (Charlottesville: University of Virginia Press, 2010), 13-14.

[122] Haines, *Horses in America*, 57.

[123] Campbell, *The Horse in Virginia*, 13.

[124] (Facts in paragraph.) Haines, *Horses in America*, 58.

[125] Dutson, *96 Horse Breeds of North America*, 22-23.

[126] Campbell, *The Horse in Virginia*, 17.

[127] Haines, *Horses in America*, 58-59.

[128] Ibid, 59.

[129] Ibid.

[130] Gruenberg, *The Hoofprints Guide*, 22.

[131] Haines, *Horses in America*, 59.

[132] Ibid.

[133] Ibid.

[134] Dutson, *96 Horse Breeds of North America*, 17.

[135] Dobie, *The Mustangs*, 22.

[136] (Facts in paragraph.) Greene, *Horses at Work*, 100.

[137] Ibid, 100-101.

138 Haines, *Horses in America*, 58.

139 Ibid.

140 John Amrhein, Jr., "La Galga the Legendary Assateague Galleon," 2014. Accessed January 11, 2020. http://thehiddengalleon.com

141 Campbell, *The Horse in Virginia*, 227.

142 Ibid.

143 (Facts in paragraph.) Campbell, 228.

144 Gruenberg, *The Hoofprints Guide*, 30.

145 Ibid, 25.

146 Ibid, 26.

147 (Facts in paragraph.) Ibid, 21-23.

148 (Facts in paragraph.) Ibid, 23-24.

149 E.K. Conant, R. Juras, and E.G. Cothran, "A Microsatellite Analysis of Five Colonial Spanish Horse Populations of the Southeastern United States," *Animal Genetics*, 43, 1 (February 2012): 53-62. https://www.ncbi.nlm.nih.gov/pubmed/22221025

150 (Facts in paragraph.) Gruenberg, *The Hoofprints Guide*, 27.

151 (Facts in paragraph.) Bonnie Gruenberg, "The Wild Horses of Cumberland Island, GA," *Wild Horse Islands*. Accessed January 11, 2020. http://www.wildhorseislands.com/cumberlandislandhorsesga.html

152 (Facts in paragraph.) Campbell, *The Horse in Virginia*, 17.

153 (Facts in paragraph.) Ibid, 17-19.

154 Ibid, 24.

155 Dutson, *96 Horse Breeds of North America*, 66.

156 Campbell, *The Horse in Virginia*, 30.

157 Dutson, *96 Horse Breeds of North America*, 66.

158 Ibid, 18.

159 Ibid.

160 New England Historical Society, "The Narragansett Pacer: The Lost Horse of the New England Colonies." Updated in 2018. Accessed January 11, 2020. http://www.newenglandhistoricalsociety.com/narragansett-pacer-lost-horse-new-england-colonies/

161 Dutson, *96 Horse Breeds of North America*, 238.

162 New England Historical Society, "The Narragansett Pacer."

163 Ibid.

164 Haines, *Horses in America*, 60.

165 Duston, *Storey's Illustrated Guide to 96 Horse Breeds of North America*, 239

166 (Facts in paragraph.) Ibid, 20.

167 Ibid, 21.

168 ("...the most efficient...") Greene, *Horses at Work*, 60. ("...required six to

eight Conestoga horses.") Duston, *Storey's Illustrated Guide to 96 Horse Breeds of North America*, 21.

[169] (Facts in paragraph.) Ibid, 21.

[170] Ibid, 22.

[171] (Facts in paragraph.) Debra Michals, "Sybil Ludington," National Women's History Museum, 2017. Accessed January 12, 2020. www.womenshistory.org/education-resources/biographies/sybil-ludington

[172] Marsha Amstel, *Sybil Ludington's Midnight Ride* (Minneapolis: Millbrook Press, 2000), 47.

[173] Karen Braschayko, "Sybil Ludington and Her Horse Star, Heroes of the American Revolution," *Equitrekking*, July 3, 2017. https://equitrekking.com/articles/entry/sybil-ludington-and-her-horse-star-heroes-of-the-american-revolution

[174] Kat Eschner, "Was There Really a Teenage, Female Paul Revere?" *Smithsonian Magazine*, April 26, 2017. https://www.smithsonianmag.com/smithsonianmag/was-there-really-teenage-female-paul-revere-180962993/

[175] (Story of Betsy Dowdy.) Kitty Griffin, *The Ride: The Legend of Betsy Dowdy* (New York: Atheneum Books for Young Readers, 2010).

[176] Michael D. Hattem, "The Historiography of the American Revolution," *Journal of the American Revolution*, August 23, 2013. https://allthingsliberty.com/2013/08/historiography-of-american-revolution/

[177] Donald N. Moran, "The Birth of the American Cavalry." Accessed September 16, 2019. http://revolutionarywararchives.org/cavalry.html

[178] Haines, *Horses in America*, 62.

[179] Ibid, 63.

[180] (Facts in paragraph.) Ibid, 62.

[181] Campbell, *The Horse in Virginia*, 32.

[182] (Facts in paragraph.) Moran, "The Birth of the American Cavalry."

[183] ("Many skilled Virginia horsemen...") Campbell, *The Horse in Virginia*, 32. ("Surprise attacks...") Haines, *Horses in America*, 63-64.

[184] Ibid, 63.

[185] Jeannette Barenger, "The Carolina Marsh Tacky—Yesterday and Today." Accessed September 16, 2019. http://www.carolinamarshtacky.com/breed-history.html

[186] Amy Crawford, "The Swamp Fox," *Smithsonian Magazine*, June 30, 2007. https://www.smithsonianmag.com/history/the-swamp-fox-157330429/

[187] Ibid.

[188] Campbell, *The Horse in Virginia*, 35.

[189] (Facts in paragraph.) Haines, *Horses in America*, 65.

190 Ibid, 66.

191 Dutson, *96 Horse Breeds of North America*, 24.

192 Ibid, 24.

193 Ibid, 25.

194 Ibid.

195 Ibid.

196 (Stolen horses and trade for 29.) Haines, *Horses in America*, 96. (Trade included promise of guns.) Merriweather Lewis, *Journals of the Lewis & Clark Expedition*, August, 17, 1805. Accessed January 5, 2020. https://lewisandclarkjournals.unl.edu/item/lc.jrn.1805-08-17#lc.jrn.1805-08-17.01

197 Merriweather Lewis, *Journals of the Lewis & Clark Expedition*, June 2, 1806. Accessed January 5, 2020. https://lewisandclarkjournals.unl.edu/item/lc.jrn.1806-06-02#lc.jrn.1806-06-02.01

198 (Facts in paragraph.) Haines, *Horses in America*, 96.

199 Ibid, 97.

200 Ibid, 102.

201 Ibid, 103.

202 Dutson, *96 Horse Breeds of North America*, 20.

203 Haines, *Horses in America*, 105.

204 Ryden, *America's Last Wild Horses*, 102.

205 Dobie, *The Mustangs*, 271.

206 Haines, *Horses in America*, 98.

207 Haines, *Horses in America*, 98.

208 Gwynne, *Empire of the Summer Moon*, 27.

209 The California Vaquero Horse Association, "The History of California's Horses." Accessed September 28, 2019. https://californiavaquero.wixsite.com/cvha/history

210 History, "California Missions." Updated August 21, 2018. https://www.history.com/topics/religion/california-missions

211 (Both population statics.) Tricia Weber, "Spanish Exploration," February 9, 2009. http://www.californias-missions.org/history.htm

212 PBS, "The Gold Rush Impact on Native Tribes." Accessed September 28, 2019. https://www.pbs.org/wgbh/americanexperience/features/goldrush-value-land/

213 California Missions Foundation, "California Indians Before, During, and After the Mission Era." Accessed September 28, 2019. http://californiamissionsfoundation.org/california-indians/

214 Haines, *Horses in America*, 107.

[215] Ibid.

[216] Ibid 107-108.

[217] (Facts in paragraph.) Ibid, 43.

[218] (Facts in paragraph.) Haines, *Horses in America*, 108.

[219] (Facts in paragraph.) Ibid, 108.

[220] Dobie, *The Mustangs*, 323.

[221] Ibid, 41.

[222] Ibid, 323.

[223] Ibid, 322-323.

[224] (Run the off cliffs...) Ibid, 323. (Slash them with a lance...) Haines, *Horses in America*, 108.

[225] Ibid, 109.

[226] History, "California Missions."

[227] The California Vaquero Horse Association, "The History of California's Horses." Accessed September 28, 2019. https://californiavaquero.wixsite.com/cvha/history

[228] Haines, *Horses in America*, 109.

[229] The California Vaquero Horse Association, "The History of California's Horses."

[230] History, "California Gold Rush," August 29, 2019. Accessed September 28, 2019. https://www.history.com/topics/westward-expansion/gold-rush-of-1849

[231] History, "California Gold Rush."

[232] Ibid.

[233] Haines, *Horses in America*, 147.

[234] Ibid, 110.

[235] History, "California Gold Rush."

[236] PBS, "The Gold Rush Impact on Native Tribes." Accessed September 28, 2019. https://www.pbs.org/wgbh/americanexperience/features/goldrush-value-land/

[237] Haines, *Horses in America*, 147.

[238] (Facts in paragraph.) Ryden, *America's Last Wild Horses*, 147.

[239] Gwynne, *Empire of the Summer Moon*, 25.

[240] Katie Nodjimbadem, "The Lesser-Known History of African-American Cowboys," *Smithsonian Magazine*, February 13, 2017. Accessed September 30, 2019. https://www.smithsonianmag.com/history/lesser-known-history-african-american-cowboys-180962144/

[241] Ryden, *America's Last Wild Horses*, 147.

[242] (Facts in paragraph.) Ibid, 148.

[243] Ibid, 148.

[244] (Facts in paragraph.) Ibid, 146.

[245] Haines, *Horses in America*, 145.

[246] Ryden, *America's Last Wild Horses*, 151.

[247] (Facts in paragraph.) Ibid, 149.

[248] (Facts in paragraph.) Ibid, 153.

[249] Dobie, *The Mustangs*, 196.

[250] Haines, *Horses in America*, 144.

[251] Dee Brown, *Bury My Heart at Wounded Knee: An Indian History of the American West.* (New York: Holt Paperbacks, 1970), 5. Reissued in 2007.

[252] Act of June 30, 1834, 4 Stat 729. Accessed October 1, 2019. https://lawfare.s3-us-west-2.amazonaws.com/staging/s3fs-public/uploads/2013/01/Act-of-June-30-1834-4-Stat-729.pdf

[253] Brown, *Bury My Heart at Wounded Knee*, 6.

[254] Ibid, 9.

[255] Cited in Julius W. Pratt, "The Origin of Manifest Destiny," *The American Historical Review*, 32, 4, July 1927, 795-798. Accessed January 5, 2020. https://www.jstor.org/stable/1837859?seq=2#metadata_info_tab_contents

[256] Gwynne, *Empire of the Summer Moon*, 112-113.

[257] Ibid, 75.

[258] (Facts in paragraph.) Ibid, 133-136.

[259] (Facts in paragraph.) Ibid, 140-141.

[260] (Facts in paragraph and the following.) Ibid, 145-149.

[261] Ibid, 159

[262] (Including facts in remainder of paragraph.) Ibid, 160.

[263] Ibid, 162.

[264] (Facts in paragraph.) Ibid, 164-165.

[265] (Facts in paragraph.) Ibid, 168-171.

[266] Ibid, 209.

[267] (Facts in paragraph.) Haines, *Horses in America*, 113-114.

[268] (Facts in paragraph.) Ibid.

[269] Ibid, 114.

[270] ("Four hundred twenty fast horses.") Ibid. ("Many California mustangs.") Dobie, *The Mustangs*, 285.

[271] Haines, *Horses in America*, 115.

[272] Ibid, 114.

[273] "Pony Express Historical Timeline." Accessed October 2, 2019. http://ponyexpress.org/pony-express-historical-timeline/

[274] Greene, *Horses at Work*, 45.

[275] Ibid, 44.

[276] (Facts in paragraph.) Greene, *Horses at Work*, 45-49.

277 Ibid, 50.

278 (Facts in paragraph.) Greene 55-57.

279 Ibid, 56.

280 Heidi Ziemer, "Two Hundred Years on the Erie Canal," New York Heritage Digital Collections, September 20, 2019. https://nyheritage.org/exhibits/two-hundred-years-erie-canal

281 David Sommerstein, "Ten Threats: The Earliest Invaders," *The Environmental Report*, October 17, 2005. https://environmentreport.org/?tag=erie-canal

282 Greene, *Horses at Work*, 64-65.

283 (Towing statistics.) Ibid, 65.

284 Ibid, 66.

285 Ibid, 68.

286 (Facts in paragraph.) Ibid, 70.

287 Lake Champlain Maritime Museum, "Horse Ferry." Accessed January 6, 2020. https://www.lcmm.org/explore/vermont-underwater-historic-preserves/horse-ferry/

288 (Facts in paragraph.) Greene, *Horses at Work*, 71-72.

289 Ibid, 74.

290 (Facts in paragraph.) Ibid, 75-76.

291 Dutson, *96 Horse Breeds of North America*, 356.

292Greene, *Horses at Work*, 104.

293 Ibid, 107.

294 Haines, *Horses in America*, 91.

295 (Justin Morgan facts.) Ralph Moody, *American Horses* (Lincoln: University of Nebraska Press, 1962).

296 (Facts in paragraph.) Dutson, *96 Horse Breeds of North America*, 180.

297 Ibid, 239-240.

298 Ibid, 240.

299 Ibid.

300 Moody, *American Horses*, 121.

301 (Facts in paragraph.) Dutson, *96 Horse Breeds of North America*, 70.

302 Moody, *American Horses*, 133.

303 Ibid.

304 Dutson, *96 Horse Breeds of North America*, 246.

305 Ibid, 66.

306 John Ehle, *Trail of Tears: The Rise and Fall of the Cherokee Nation* (New York: Anchor Books, 1988), 222.

307 Ibid.

308 Ibid, 323.

309 Ibid, 295.

310 Ibid, 296.

311 (Facts in paragraph.) History, "Trail of Tears," September 30, 2019.
https://www.history.com/topics/native-american-history/trail-of-tears

312 Phillip Sponenberg, "Return to Freedom's Choctaw Herd," Return to
Freedom. Accessed October 16, 2019. https://returntofreedom.org/what-we-
do/sanctuary/our-horses/choctaw-herd/

313 Deborah Donohue, "Choctaw Horse," *Cowgirl Magazine*, April 3, 2017.
https://cowgirlmagazine.com/choctaw-horse/

314 (Description of raid.) Bruce J. Dinges, "America's Civil War: Colonel
Benjamin Grierson's Raid in 1863," History Net. Accessed June 10, 2019.
https://www.historynet.com/americas-civil-war-colonel-benjamin-griersons-
cavalry-raid-in-1863.htm

315 Haines, *Horses in America*, 117.

316 Greene, *Horses at Work*, 121.

317 Ibid, 121.

318 Ibid, 163.

319 Eric J. Wittenberg, "The Loyal Steeds: Horses in the Civil War," Civil War
Cavalry, December 25, 2012. http://civilwarcavalry.com/?p=3521

320 Greene, *Horses at Work*, 122.

321 Ibid, 124.

322 Campbell, *The Horse in Virginia*, 69.

323 Greene, *Horses at Work*, 123.

324 Haines, *Horses in America*, 124.

325 Greene, *Horses at Work*, 123.

326 Wittenberg, "The Loyal Steeds: Horses in the Civil War."

327 American Battlefield Trust, "Civil War Facts." Accessed June 10, 2019.
https://www.battlefields.org/learn/articles/civil-war-facts accessed

328 (Glanders information.) Greene, *Horses at Work*, 143.

329 Ibid, 150-151.

330 Ibid, 139.

331 Ibid.

332 Haines, *Horses in America,* 117.

333 Campbell, *The Horse in Virginia*, 69.

334 Ibid, 74.

335 "Surrender Documents," April 9-10, 1865. Accessed January 7, 2020.
https://www.nps.gov/apco/learn/education/surrender-documents.htm

336 Greene, *Horses at Work*, 120.

337 Lillian Schlissel, *Black Frontiers: A History of African American Heroes of the
Old West*. (New York: Aladdin Paperbacks, 1995), 54-55.

[338] Philipps, *Wild Horse Country*, 72.

[339] Ryden, *America's Last Wild Horses*, 146.

[340] Ibid, 123.

[341] Ibid, 155-158.

[342] Philipps, *Wild Horse Country*, 70-71.

[343] Ryden, *America's Last Wild Horses*, 156.

[344] Ibid, 146.

[345] Ibid, 153.

[346] Sarah Maslin Nir, "Restoring Black Cowboys to the Range," *The New York Times*, September 14, 2019.
https://www.nytimes.com/2019/09/14/travel/black-cowboy-museum-texas.html?action=click&module=Features&pgtype=Homepage

[347] Nodjimbadem, "The Lesser-Known History of African-American Cowboys."

[348] (Facts in paragraph.) Schlissel, *Black Frontiers*, 30, 31, 36.

[349] Nodjimbadem, "The Lesser-Known History of African-American Cowboys."

[350] Schlissel, *Black Frontiers*, 30.

[351] Joyce Gibson Roach, *The Cowgirls* (University of North Texas Press, 1977).

[352] (Facts in paragraph.) Ibid, 83-84.

[353] Ibid, 83-85.

[354] Ryden, *America's Last Wild Horses*, 155.

[355] Haines, *Horses in America*, 163.

[356] Philipps, *Wild Horse Country*, 79.

[357] Haines, *Horses in America*, 166.

[358] Ryden, *America's Last Wild Horses*, 159, 161.

[359] Haines, *Horses in America*, 166.

[360] Ibid, 166.

[361] Recounted by Dobie, *The Mustangs*, 216-217. Original citation: N. Howard (Jack) Thorpe, *Partner of the Wind* (In collaboration with Neil M. Clark) (Caldwell: Caxton Printers, Ltd., 1945).

[362] Dobie, *The Mustangs*, 196.

[363] Ibid, 224.

[364] (Facts in paragraph.) Ibid, 221-231.

[365] (Both dollar figures.) Philipps, *Wild Horse Country*, 76.

[366] Dobie, *The Mustangs*, 316.

[367] Ryden, *America's Last Wild Horses*, 190.

[368] Dobie, *The Mustangs*, 139.

[369] Ibid.

[370] Gwynne, *Empire of the Summer Moon*, 222.

[371] Ibid, 213.

[372] Ibid, 222.

373 Ibid, 260.
374 (Remainders of facts in paragraph.) Ryden, *America's Last Wild Horses*, 174-175.
375 Gwynne, *Empire of the Summer Moon*, 260.
376 Ryden, *America's Last Wild Horses*, 167.
377 (Facts in remainder of paragraph.) Andrew C. Isenberg, "Toward a Policy of Destruction: Buffaloes, Law, and the Market, 1803-83," *Greater Plains Quarterly*, 12, 4 (Fall 1992): 227-241.
https://www.jstor.org/stable/23531659?seq=1#metadata_info_tab_contents
378 (Remainder of facts in paragraph.) Gwynne, *Empire of the Summer Moon*, 231-233.
379 FrontierTexas, "Meet the Legends: Ranald Mackenize." Accessed March 29, 2020. http://frontiertexas.com/biographies/ranald-mackenzie
380 Gwynne, *Empire of the Summer Moon*, 242-248.
381 H. Allen Anderson, "Battle of Blanco Canyon," *Handbook of Texas Online*. Updated June 12, 2010.
https://tshaonline.org/handbook/online/articles/qfb02
382 (Facts in paragraph.) Gwynne, *Empire of the Summer Moon*, 255-256.
383 Ibid, 256.
384 Ibid, 257.
385 (Facts in rest of paragraph.) Ibid, 272-274.
386 Ibid, 276.
387 (Facts in paragraph.) Ibid, 280-282.
388 Thomas F. Schilz, "Battle of Palo Duro Canyon," *Handbook of Texas Online*. Updated June 15, 2010.
https://tshaonline.org/handbook/online/articles/btp03
389 Gwynne, *Empire of the Summer Moon*, 283.
390 Nerburn, *Chief Joseph & the Flight of the Nez Perce*, 39.
391 Ibid, 40.
392 (Facts in paragraph.) Ibid.
393 (Remainder of facts in paragraph.) Ibid, 52.
394 Ibid, 54.
395 Ibid, 55.
396 Jermone A. Greene, *Nez Perce Summer 1877* (Helena: Montana Historical Society Press, 2000).
https://www.nps.gov/parkhistory/online_books/nepe/greene/chap1.htm
397 Nerburn, *Chief Joseph & the Flight of the Nez Perce*, 87.
398 Ibid, 88.
399 (Facts in paragraph.) Ibid, 90.
400 Ibid, xi.

[401] Ibid, 119.

[402] Ibid, xi.

[403] Ibid, 278.

[404] Haines, *Horses in America*, 158.

[405] Nerburn, *Chief Joseph & the Flight of the Nez Perce*, 266-268.

[406] Ibid, 268.

[407] Ibid, 296.

[408] Ibid, 298.

[409] Appaloosa Museum, "Recent History of the Appaloosa." Accessed October 12, 2019. https://www.appaloosamuseum.org/recent-history-of-the-appaloosa/

[410] Appaloosa Museum, "History of APHC." Accessed October 12, 2019. http://www.appaloosamuseum.com/history-of-APHC

[411] Deborah Donohue, "The Nez Perce Horse," *Cowgirl Magazine*, July 27, 2017. https://cowgirlmagazine.com/nez-perce-horse/

[412] (Facts in paragraph.) Ibid.

[413] (Facts in paragraph.) Brown, *Bury My Heart at Wounded Knee*, 276.

[414] Ibid, 279.

[415] Ibid, 284.

[416] Ibid, 285.

[417] (Facts in paragraph) Thomas Powers, "How the Battle of Little Bighorn was Won," *Smithsonian Magazine*, November 2010. (Adapted from *The Killing of Crazy Horse*.) https://www.smithsonianmag.com/history/how-the-battle-of-little-bighorn-was-won-63880188/

[418] Ibid, 297-298.

[419] Ibid, 300.

[420] Ibid.

[421] Ibid, 306.

[422] Ibid, 308.

[423] Frederick Jackson Turner, *The Frontier in American History*, 1919. Accessed January 7, 2020. https://courses.lumenlearning.com/ushistory2os/chapter/primary-source-frederick-jackson-turner-significance-of-the-frontier-in-american-history-1893/

[424] Mary Stewart Atwell, "Searching Out the Hidden Stories of South Carolina's Hidden Culture," *The New York Times*, April 15, 2019. https://www.nytimes.com/2019/04/15/travel/south-carolina-gullah-geechee-low-country.html?action=click&module=Top%20Stories&pgtype=Homepage

[425] Henry Louis Gates, Jr., "The Truth Behind 40 Acres and a Mule," PBS. Accessed October 14, 2019.

https://www.pbs.org/wnet/african-americans-many-rivers-to-cross/history/the-truth-behind-40-acres-and-a-mule/
426 Ibid.
427 "The Marsh Tacky Horse," *Gullah Magazine*. Accessed October 14, 2019. http://gullahcelebration.com/gullah-magazine/culture/the-marsh-tacky-horse
428 Ibid.
429 Ibid.
430 Campbell, *The Horse in Virginia*, 102.
431 Maryjean Wall, *How Kentucky Became Southern: A Tale of Outlaws, Horse Thieves, Gamblers, and Breeders* (Lexington: The University Press of Kentucky, 2010), 109-110.
432 Campbell, *The Horse in Virginia*, 103.
433 Wall, *How Kentucky Became Southern*, 113.
434 (Facts in paragraph.) Greene, *Horses at Work*, 174.
435 (Facts in paragraph.) Ibid, 166.
436 Ibid.
437 Ibid, 173-174.
438 Ibid, 177.
439 (Facts in this paragraph and the following.) Ibid, 178, 184.
440 (Facts in remainders of paragraph.) Philipps, *Wild Horse* Country, 72-73.
441 (Facts in paragraph.) Greene, *Horses at Work*, 172-175.
442 Ibid, 175.
443 (Facts in paragraph.) Ibid, 189-192.
444 Ibid, 199.
445 Haines, *Horses in* America, 170.
446 Sean Kheraj, "The Great Epizootic of 1872–73: Networks of Animal Disease in North American Urban Environments," *Environmental History*, 23, 3 (July 2018): 495–52. https://academic.oup.com/envhis/article/23/3/495/4985859
447 Greene, *Horses at Work*, 168.
448 Ibid.
449 Kheraj, "The Great Epizootic of 1872–73: Networks of Animal Disease in North American Urban Environments."
450 (Facts in paragraph.) Greene, *Horses at Work*, 231.
451 (Facts in paragraph.) Ibid, 174.
452 Ibid, 241.
453 Ibid, 242.
454 (Facts in paragraph.) Ibid, 186-188.
455 Wall, *How Kentucky Became Southern,* 3.
456 (Facts in paragraph.) Ibid, 28-34.
457 Ibid, 21.

[458] Ibid, 62.

[459] Gary Adelman and Mary Bays Woodside, "A House Divided: Civil War in Kentucky," American Battlefield Trust. Updated December 9, 2019. https://www.battlefields.org/learn/articles/house-divided-civil-war-kentucky

[460] Wall, *How Kentucky Became Southern,* 63.

[461] Ibid, 14-16.

[462] (Facts in paragraph.) Ibid, 28.

[463] Ibid, 77.

[464] Ibid, 58

[465] Ibid, 67.

[466] Ibid, 3.

[467] Donna M. Campbell, "Plantation Tradition in Local Color Fiction," *Literary Movements*. Last modified September 7, 2015. https://public.wsu.edu/~campbelld/amlit/plant.htm

[468] Wall, *How Kentucky Became Southern*, 202.

[469] (Facts in paragraph.) Ibid, 205.

[470] Ibid, 160.

[471] Ibid, 219.

[472] Holly Wiemers, "Kentucky's Equine Industry Had $3 Billion Impact," College of Agriculture, Food and Environment. Accessed October 16, 2019. https://equine.ca.uky.edu/news-story/kentuckys-equine-industry-has-3-billion-economic-impact

[473] Greene, *Horses at Work*, 268.

[474] Ibid, 253-254.

[475] Haines, *Horses in America*, 189.

[476] Ibid, 189.

[477] Greene, *Horses at Work*, 266.

[478] Philipps, *Wild Horse Country*, 81.

[479] Greene, *Horses at Work*, 269.

[480] Ibid, 271.

[481] Ibid, 272.

[482] U.S. Department of Commerce and Bureau of the Census, *The Farm Horse,* (Washington: United States Government Printing Office, 1933). Accessed January 8, 2020. https://www2.census.gov/library/publications/decennial/1930/agriculture-farm-horse/1930sr-farm-horse.pdf

[483] Greene, *Horses at Work*, 267.

[484] Ibid, 188.

[485] The International Museum of the Horse, "1900—The Horse in Transition." Accessed July 19, 2019. http://imh.org/legacy-of-the-horse/the-horse-in-world-war-i-1914-1918/

[486] Ryden, *America's Last Wild Horses*, 200.

[487] The International Museum of the Horse, "1900—The Horse in Transition."

[488] Philipps, *Wild Horse Country*, 81.

[489] Ibid, 60.

[490] Ibid, 83-84.

[491] Ibid, 85.

[492] (Facts in the rest of paragraph.) Ibid, 131-133.

[493] Elizabeth Letts, *The Perfect Horse: The Daring U.S. Mission to Rescue the Priceless Stallions Kidnapped by the Nazis* (New York: Ballantine Books, 2016), 104.

[494] (Facts in paragraph and the following.) Ibid, 112-114.

[495] ("The Twenty-Sixth Cavalry...) Rickey Robertson, "Horse Tales of the U.S. Cavalry in Louisiana—1941," Center for Regional Heritage Network. Accessed January 8, 2020. http://www.sfasu.edu/heritagecenter/9246.asp . (That is, until 2001...) Dwight Jon Zimmerman, "21st Century Horse Soldiers—Special Operations Forces and Operation Enduring Freedom," *Defense Media Network*, September 11, 2011. https://www.defensemedianetwork.com/stories/operation-enduring-freedom-the-first-49-days-4/

[496] Robertson, "Horse Tales of the U.S. Cavalry in Louisiana—1941."

[497] (Facts in paragraph.) Merrit Clifton, "The Fort Polk Horses: Last Stand of the U.S. Cavalry," Animals 24-7. Accessed January 9, 2020. https://www.animals24-7.org/2018/08/17/the-fort-polk-horses-last-stand-of-the-u-s-cavalry/

[498] *Glimpses*, "The Marsh Tacky in History," February 2010. http://marshtacky.info/mt/wp-content/uploads/2016/01/Glimpses.pdf

[499] The Seagoing Cowboys, "World War II Ships Re-Purposed as Livestock Carriers," August 28, 2015. https://seagoingcowboysblog.wordpress.com/2015/08/28/world-war-ii-ships-re-purposed-as-livestock-carriers/

[500] Wessels Living History Farm, "UNRRA and Private Relief." Accessed January 9, 2020. https://livinghistoryfarm.org/farmingin the40s/money_06.html

[501] (Facts in paragraph.) Letts, *The Perfect Horse*, 24 and 61-62.

[502] Ibid, 42.

[503] Ibid, 61.

[504] (Facts in remainder of paragraph.) Ibid, 98.

[505] (Facts in paragraph.) Ibid, 158-162.

[506] (Facts in remainder of paragraph.) Ibid, 161-179.

[507] Ibid, 166.

[508] Ibid, 193.

[509] Ibid, 203.

[510] (Facts in paragraph.) Ibid, 223-232.

[511] Ibid, 236-238.

[512] Ibid.

[513] Ibid, 222.

[514] Ibid, 236 and 244.

[515] Ibid, 294.

[516] (Facts in paragraph.) Philipps, *Wild Horse Country*, 130.

[517] Ibid, 133.

[518] Ibid, 139.

[519] Ibid, 140-142.

[520] Ryden, *America's Last Wild Horses*, 224.

[521] Philipps, *Wild Horse Country*, 147.

[522] Ryden, *America's Last Wild Horses*, 238-239.

[523] (Facts in paragraph.) Ibid, 239.

[524] (Facts in paragraph.) Ibid, 245-248.

[525] Ibid, 256.

[526] (Facts in paragraph.) Ibid, 257-261.

[527] Ibid, 261.

[528] Gus E. Cothran, "Genetic Analysis of the Pryor Mountains Wild Horse Range, MT," August 22, 2013. https://static1.squarespace.com/static/59f8c99ff09ca4e7c237d467/t/5a0f09e824a6944508b81b82/1510935017512/PryorMNTS2012GeneticReport.pdf

[529] *The Wild Free-Roaming Horses and Burros Act of 1971*. Accessed January 9, 2020. https://www.blm.gov/or/regulations/files/whbact_1971.pdf

[530] Bureau of Land Management Wild Horse and Burro Program, "Myths and Facts." Accessed January 9, 2020. https://www.blm.gov/programs/wild-horse-and-burro/about-the-program/myths-and-facts

[531] U.S. Fish and Wildlife Service, "BLM, FWS Agreement Enhances Management of Wild and Feral Horses and Burros," News Release, September 2, 2010. https://www.fws.gov/news/ShowNews.cfm?ref=blm-fws-agreement-enhances-management-of-wild-and-feral-horses-and-burros&_ID=1345

[532] U.S. Fish and Wildlife Service, "Feral Horse & Burro Management." Accessed January 9, 2020. https://www.fws.gov/sheldonhartmtn/pdf/Feral%20Horse.pdf

[533] (Facts in paragraph) Jenny Powers, "Management of Feral Horses in the National Park Service," *The American Mustang*, AAEP Proceedings, 60 (2014): 433-436. https://www.horsesforlife.org/uploads/1/0/5/8/10585042/fertility_control_wild_horses_-_powers.pdf

534 Kay Frydenborg, *Wild Horse Scientists* (Boston: Houghton Mifflin Harcourt, 2012), 47, 51.

535 National Academy of Sciences, *Using Science to Improve the BLM Wild Horse and Burro Program: A Way Forward*, 2013. Accessed July 30, 2019. http://dels.nas.edu/resources/static-assets/materials-based-on-reports/reports-in-brief/wild-horses-report-brief-final.pdf

536 ("88,000") Steve Tryon, "Wild Horses and Burros," Bureau of Land Management, July 16, 2019. Accessed January 24, 2020. https://www.doi.gov/ocl/wild-horses-and-burros-0 . ("27,000") Bureau of Land Management, "Bureau of Land Management Wild Horse and Burro Program." Accessed October 21, 2019. https://www.blm.gov/sites/blm.gov/files/wildhorse_2019infographic_52119_compressed_v2.pdf . ("Insisting it could get...") Philipps, *Wild Horse Country*, 163, 220.

537 Bureau of Land Management, "Bureau of Land Management Wild Horse and Burro Program."

538 (Facts in paragraph.) National Academy of Sciences, *Using Science to Improve the BLM Wild Horse and Burro Program: A Way Forward.*

539 Jay F. Kirkpatrick and John W. Turner, Jr., "Compensatory Reproduction in Feral Horses," *The Journal of Wildlife Management*, 55, 4 (October 1991): 649-652. https://www.jstor.org/stable/3809514?seq=1

540 Study cited in Frydenborg, *Wild Horse Scientists*, 51.

541 National Academy of Sciences, *Using Science to Improve the BLM Wild Horse and Burro Program: A Way Forward.*

542 Philipps, *Wild Horse Country*, 275.

543 Ibid, 277.

544 Ibid, 246.

545 ASPCA, "ASPCA urges public support for the American Horse Slaughter Prevention Act," February 1, 2012. https://www.prnewswire.com/news-releases/aspca-research-confirms-americans-strongly-oppose-slaughter-of-horses-for-human-consumption-138494089.html

546 Scott Streater, "Budget Calls To Euthanize Animals it Can't Adopt Out," American Wild Horse Campaign. Accessed October 22, 2019. https://americanwildhorsecampaign.org/media/budget-calls-blm-euthanize-animals-it-cant-adopt-out

547 Philipps, *Wild Horse Country*, 178.

548 Ibid, 183.

549 Ibid, 185.

550 Martha Mendoza, "Trails End for Horses: Slaughter," *LA Times*, January 5, 1997.

551 Philipps, *Wild Horse Country*.

552 Frydenborg, *Wild Horse Scientists*, 14.

553 (Facts in paragraph ad the following.) Ibid, 15-16.

554 (Facts in remainder of paragraph and the following.) Ibid, 47-51.

555 Philipps, *Wild Horse Country*, 255.

556 (Facts in paragraph.) American Wild Horse Campaign, "The Issue." Accessed July 31, 2019. https://americanwildhorsecampaign.org/issue

557 National Academy of Sciences, *Using Science to Improve the BLM Wild Horse and Burro Program: A Way Forward*.

558 Karen Brulliard. "The Battle Over Wild Horses," *The Washington Post*, September 18, 2019.
https://www.washingtonpost.com/science/2019/09/18/wild-horses-have-long-kicked-up-controversy-now-foes-say-they-have-solution/?arc404=true

559 Ibid.

560 American Wild Horse Campaign, "Euthansia: A No-Go for Wild Horses." Accessed July 31, 2019. https://americanwildhorsecampaign.org/media/blm-chief-euthanasia-no-go-wild-horses

561 (Description of plan.) *The Path Forward for Management of BLM's Wild Horses & Burros*, 2019. Accessed January 24, 2020.
https://www.energy.senate.gov/public/index.cfm/files/serve?File_id=0869B0 2B-E9C5-4F0B-9AE8-9A8A1C85293E

562 Brulliard, "The Battle Over Wild Horses."

563 Karin Brulliard, "Spending Bill Includes Funding for Controversial Wild Horse Proposal," *The Washington Post*, December 21, 2019. Accessed January 24, 2020. https://www.washingtonpost.com/science/2019/12/21/spending-bill-includes-funding-controversial-wild-horse-proposal/

564 ($25 billion industry) Reuters, "Horse Racing Fading in Revenue, Popularity," *Newsweek*, May 8, 2016. https://www.newsweek.com/horse-racing-fading-revenue-popularity-457123 (1.2 million horses) American Horse Council, *Economic Impact of the United States Horse Industry*. Accessed August 5, 2019.

565 Joe Drape, "At Santa Anita, Horse Must Now Be Evaluated by a Safety Team Before Racing," *New York Times*, June 12, 2019.
https://www.nytimes.com/2019/06/12/sports/santa-anita-park-horses-deaths.html?searchResultPosition=2

566 The Jockey Club, *Vision 2025: To Prosper, Horse Racing Needs Comprehensive Reform*, March 28, 2019, 5.
http://jockeyclub.com/pdfs/vision_2025.pdf

567 Walt Bogdanich, Joe Drape, and Rebecca R. Ruiz, "At the Track, Racing Economics Collide with Veterinarians' Oath," *The New York Times*, September

22, 2012. https://www.nytimes.com/2012/09/22/us/at-the-track-racing-economics-collide-with-veterinarians-oath.html

568 The Jockey Club, *Vision 2025: To Prosper, Horse Racing Needs Comprehensive Reform.*

569 Joe Drape, "PETA Accuses Two Trainers of Cruelty to Horses," *The New York Times*, March 19, 2014. https://www.nytimes.com/2014/03/20/sports/peta-accuses-two-trainers-of-cruelty-to-horses.html

570 Tom Goldman, "23 Thoroughbred Deaths Force Anita to Change. Will the Racing Industry Follow?" *The New York Times*, April 10, 2019. https://www.npr.org/2019/04/10/711726633/23-thoroughbred-deaths-force-santa-anita-to-change-will-the-racing-industry-foll

571 (Facts in paragraph.) Joe Drape and Corina Knoll, "Why So Many Horses Have Died at Santa Anita," *The New York Times*, June 26, 2019. https://www.nytimes.com/2019/06/26/sports/santa-anita-horse-deaths.html?action=click&module=RelatedLinks&pgtype=Article

572 Jacey Fortin and Kevin Draper, "California Governor, Citing 29 Horse Deaths, Calls for Hold on Racing at Santa Anita Track," *The New York Times*, June 11, 2019. https://www.nytimes.com/2019/06/11/sports/horse-racing/santa-anita-close-track.html?searchResultPosition=1

573 Joe Drape, "At Santa Anita, Horse Must Now Be Evaluated by a Safety Team Before Racing."

574 Los Angeles County District Attorney's Office, *Santa Anita Task Force Report of Investigation*, December 2019. http://da.lacounty.gov/sites/default/files/pdf/LADA-Santa-Anita-Task-Force-Report.pdf

575 Neil Vigdor, "3 Horses Die in 3 Days at Santa Anita, Prompting Free Criticism of Racetrack," *The New York Times*, January 20, 2020. https://www.nytimes.com/2020/01/20/sports/Horse-deaths-euthanized-Santa-Anita.html?searchResultPosition=1

576 (Facts in paragraph.) Joe Drape, *"Justify Failed a Drug Test Before Winning the Triple Crown," The New York Times,* September 11, 2019. https://www.nytimes.com/2019/09/11/sports/horse-racing/justify-drug-test-triple-crown-kentucky-derby.html?searchResultPosition=2

577 Tim Layden, "The Story Behind Justifies Positive Drug Test," NBC Sports, September 25, 2019. https://sports.nbcsports.com/2019/09/25/story-behind-justify-positive-drug-test/

578 (Facts in paragraph.) Benjamin Weiser and Joe Drape, "More than Two Dozen Charged in Horse Racing Doping Scheme," *The New York Times*, March 9, 2020. https://www.nytimes.com/2020/03/09/sports/horse-racing-doping.html?searchResultPosition=1

[579] (Facts in paragraph) Tom Sullivan, "Even a Triple Crown Won't End the Problems Facing Horse Racing," *Louisville Courier Journal*, June 6, 2018. https://www.courier-journal.com/story/sports/2018/06/06/triple-crown-wont-stop-problems-facing-horse-racing/674375002/

[580] BloodHorse Staff, "Reps. Barr, Tonko Reintroduce Horseracing Integrity Act," *Blood Horse*, March 14, 2019. https://www.bloodhorse.com/horse-racing/articles/232536/reps-barr-tonko-reintroduce-horseracing-integrity-act

[581] Ibid.

[582] American Horse Council, *Economic Impact of the United States Horse Industry*.

[583] ($122 billion per year.) Ibid. ("...a greater impact on the economy than the motion picture industry") Equo, "The Horse Industry by the Numbers," January 16, 2007. https://www.ridewithequo.com/blog/the-horse-industry-by-the-numbers

[584] ("7.2 million horses") American Horse Council, *Economic Impact of the United States Horse Industry*. ("Twenty-five million horses") Greene, *Horses at Work*, 166.

[585] American Horse Council, *Economic Impact of the United States Horse Industry*.

[586] Ibid

[587] Equo, "The Horse Industry by the Numbers."

[588] Ibid.

[589] CORRAL Riding Academy. Accessed August 6, 2019. https://corralriding.org

[590] War Horses for Heroes. Accessed January 11, 2020. https://warhorsesforheroes.org

[591] Steven Kurtuz, "Wild Horses and the Inmates Who 'Gentle' Them," *New York Times*, October 5, 2017. https://www.nytimes.com/2017/10/05/fashion/mens-style/prison-horses-rehabilitation-gentling.html?partner=IFTTT

[592] Karen Bachi, "An Equine-Facilitated Prison-Based Program: Human-Horse Relations and Effects on Inmate Emotions and Behaviors," *CUNY Academic Works*, 2015. Accessed August 6, 2019. https://academicworks.cuny.edu/cgi/viewcontent.cgi?article=1161&context=gc_etds

[593] William Benda, Nancy H. McGibbon, and Kathryn L. Grant, "Improvements in Muscle Symmetry in Children with Cerebral Palsy After Equine-Assisted Therapy (Hippotherapy)," *The Journal of Alternative and Complementary Medicine*, December 2003. https://www.liebertpub.com/doi/abs/10.1089/107555303771952163

ACKNOWLEDGEMENTS

I owe thanks to many people without whom I never could have transformed this project into a polished book.

I was lucky to be raised in a family of critical thinkers, debaters, writers, and editors. This book is truly a family affair. Thank you to my editor, who happens to look like she could be my sister. Because she is my sister, Erin Connors. Erin took her pen to an early draft of the book and showed me where additional anecdotes and primary sources would elevate the story and why I should turn my original conclusion into my introduction. She read chapter revision upon chapter revision, helping me tighten both my positions and prose. And she did all this while writing her own book.

Thank you to my mom, Connie Connors, for serving as my meticulous copyeditor and for calling me out when I overreached in my assessments and when the font in which my name appeared on the cover just wasn't pretty enough. Thank you to my dad, Tim Connors, for serving as my science advisor. No one should be subjected to a book that makes evolution sound like magic, and my dad helped to make sure I wouldn't be the purveyor of badly written science.

Thank you to my sisters, Megan Swan and Mari Melby, who, despite busy work schedules and family obligations, took the time not only to advise me every time I called to lament about how I would never finish the book and how I would never figure out how to launch and market it, but also to proofread it.

Thank you to Robert Spannring for taking a chance on an indie author and creating the incredible, original art that brings this story to life. Collaborating with Robert has been a joy.

Thank you to my aunt, Bridget Bean, for sharing her passion for science with me for my whole life and always cheering on my adventures. (And, of course, for introducing me to Robert.)

Thank you to my loyal journalism school friend, Kirsten Beattie, who supported this idea from the beginning and offered her proofreading skills to make sure the finished product was up to snuff.

Thank you to my first friend in the world, my original backyard neighbor, Jen DuBos, whose lifelong love of horses is contagious. Jen flew out to Montana and Wyoming with me last year to observe wild horses, patiently answered even my most inane equine questions, and read a draft of the book to make sure I wasn't saying anything too absurd.

Thank you to Steve and Nancy Cerroni of PryorWild for providing an outstanding tour of the wild horses of the Pryor Mountains in Montana and Wyoming, hosting Jen and me on their gorgeous ranch, and generously sharing their knowledge of wild and domestic horses with us. They also allowed me to use photos of one of their Quarter Horses in Chapter Two.

Thank you to riding instructor Kelly Hunter, who has been the most caring, conscientious, and patient teacher for my girls (and for me, because I never stop asking her horse questions). We are so fortunate to have Kelly in our lives. She has encouraged this project from the moment she heard about it and was kind enough to serve as a reader.

Thank you to Piper Jones for sharing her beloved equine companions and her horsemanship knowledge with my girls. We love being part of the Rolling Hills Stables community.

Thank you to my wonderful friend, the digital marketing guru Dana Cassell, for helping me to create a marketing strategy for this book. But really, I've been leaning on Dana as an advisor in all aspects of life and parenting since we met during our oldest kids' first year of preschool. Helping to launch this book is just one more line item to add to a long list of thank yous I owe Dana.

There are many other family and friends out there who have listened to my constant talk about this project and supported my efforts to turn this book into a reality for the last two years. I appreciate you to no end.

Thank you to Jeff, the husband of my dreams. I highly recommend meeting your partner in journalism school, because you inherently gain an editor-for-life along with your marriage license. Jeff is always my first reader, my last reader, and the person who listens to my endless obsessing over projects. He also manages to find the time and energy to be a true co-parent despite his demanding work and travel schedule. While I've been working on this book, he has happily taken the girls on Saturday adventures just about every weekend to allow me time to write. I could not be fulfilling my ambitions of being a stay-at-home mom, writer, photographer, and home educator without his genuine support.

And lastly, I owe the biggest thank you to my three daughters, Cricket, Nora, and Piper. My work is always about trying to make the world a better place for them and their peers in some small way, or trying to make them more thoughtful citizens of the world. I wrote this book for them, to celebrate their passion for horses and also to show them that seeking out unique perspectives will always enrich their own. I wanted to show them we always need to do better and try harder to understand the world around us, including the people, animals, and ecosystems in it. That said, the girls had to sacrifice a lot of my time and attention over the last couple years as I pieced this book together, and yet they supported me every day of this project. There is nothing more motivating than hearing your children say they are proud of you.

ABOUT THE ILLUSTRATOR

Robert Spannring has been a professional painter and illustrator for fifty years. Born and raised in Livingston, Montana, he grew up sneaking away from school to draw the landscape and its wildlife and explore what might be around the next corner. He began his career as an illustrator and watercolorist. He was the Artist-In-Residence at the Lake Hotel in Yellowstone National Park, where his work is exhibited throughout the hotel. He also worked with Jack Horner at the Museum of the Rockies, providing the interpretive dinosaur illustrations for the Special Museum Exhibits. Spannring has done illustration work for the Greater Yellowstone Coalition (of which he is a founding member), Orvis, Winston Rods, Defenders of Wildlife, the U.S. Forest Service, the Montana Department of Fish, Wildlife and Parks, the Boone and Crockett Club, Montana Audubon Council, Roberts Reinhart Publishers, and Sasquatch Books, among others. Spannring added oil and plein air painting to his artistic repertoire and is now a member of the distinguished Montana Painters Alliance, the Impressionist Society, and the Oil Painters of America. His work has been selected for exhibits and auctions by the Charles M. Russell Museum, the Northwest Museum of Arts and Culture, the Yellowstone Art Museum, the Missoula Art Museum, the Paris Gibson Square Museum of Art, the Hockaday Museum of Art, the WaterWorks Art Museum, and galleries throughout the West. Spannring's work is also in private collections throughout the U.S. You can find him online at:

www.robertspannring.com   IG: @spannringfineart

robert@robertspannring.com   FB: Robert Spannring

ABOUT THE AUTHOR

Minnesota native Julia Soplop is a lifelong writer and photographer. A fascination with documenting animal behavior has led her around the globe, from conducting field research on lemurs in Madagascar, to swimming with sea lions in the Galápagos Islands, to studying sea turtles on the Outer Banks. So, naturally, when her daughters started riding lessons, it was the horses' behavior that first drew Soplop in. Curiosity piqued, she dove into her kids' horse books. What began innocently as light reading rapidly escalated into amassing a collection of horse literature; dashing around the country to photograph wild horses; and, ultimately, writing this book. Soplop's work has appeared in numerous publications, including *National Geographic Magazine Online*, *Design Mom*, *Skiing*, *The Summit Daily News*, *Vail Daily*, *Duke Magazine*, and *The Chapel Hill News*. She is also the author of *Documenting Your World Through Photography*. Soplop has a bachelor's in French from Duke University and a master's in Medical Journalism from the University of North Carolina at Chapel Hill. She lives with her husband, their three girls, and a spritely hedgehog in the woods outside of Chapel Hill. You can find her online at:

www.juliasoplop.com   IG: @juliasoplop

julia@juliasoplop.com   FB: juliasoplopauthor

A NOTE TO THE READER

Dear Reader,

I made the decision to publish *Equus Rising* with a small, independent press to retain editorial control and the ability to hire the illustrator of my choice. (Thankfully I got my wish, and artist Robert Spannring agreed to work with me.) It means I don't have a massive publishing house peddling the book for me, though. Instead, my hope of success lies in your hands. If you enjoyed *Equus Rising*, I encourage you to leave starred ratings and reviews on Amazon and Goodreads. The more positive reviews the book receives, the more visible it becomes to future readers. Would you please take a moment to help *Equus Rising* stand out?

Many thanks,
Julia

CPSIA information can be obtained
at www.ICGtesting.com
Printed in the USA
FSHW020640310720
71942FS